I0473292

Typhon:
Reaper of Worlds

Olaf H. Hage III

Typhon: Reaper of Worlds

White Cottage Publishing Company
201 Raton Street, Suite #112
Trinidad, Colorado 81082

Typhon: Reaper of Worlds

Copyright © 2019 by the Olaf H. Hage III Literary Trust, all rights reserved. This book is protected under the copyright laws of the United States of America and applicable international copyright treaties. No portion of this book may be reproduced, stored in a retrieval system, or transmitted in any form or by any other means without prior express written permission of the management of White Cottage Publishing Company. Reviewers and other writers are hereby granted permission under the "Fair Use Doctrine" of the United States Copyright Law, Title 17 U.S.C. § 107 to use brief quotations in articles and reviews, provided proper credit is given to the Author. Any paid e-download of this book results only in a license to the recipient of the download and may not be shared with or forwarded to any other reader.

Cover, Interior Design, and editing by Tom Mack, MBA
White Cottage Publishing Company, Trinidad, Colorado
Website: http://whitecottagepublishing.com
Author Website: http://olafhage.com
Cover image licensed to the Author by Adobe Images
ISBN: 978-1072564805

Printed in the United States of America for Worldwide Distribution

Dedication

For Elizabeth, only you can know how much I miss you.

Table of Contents

Foreword

As we get closer to the time of the end, countless students of the Bible continue to study the prophecies, trying to piece together the end-time chain of events. In efforts to do this, I have seen people put together elaborate charts, spreadsheet calculations that defy the mind, and we have also seen voluminous manuscripts which attempt to figure out the progression of events that make up the end times. So far, these massive compilations have proven to be wrong. They have been wrong because they have failed to understand the "Seventh Seal Judgment" and the "Seven Trumpets" of Revelation 8-9.

The sad reality is that most end-time researchers wrongly accept assumptions which they believe are biblical fact, but which, when you study the history of these assumptions, you discover they come from untoward sources. Many of these doctrines come from gnostic writers trying to sell their ideas into the Christian community and make some money. Other ideas wandered into the Christian community from cults that teach a considerable number of erroneous doctrines. When ideas wander into the Christian community from various cults or false religious, the Bible warns us not to accept their ideas.[1]

---

[1]Deuteronomy 13:1-3, I Timothy 1:10, II Timothy 4:3-4, II Peter 2:1-2

In legal doctrine, they call evidence from a tainted source: "the fruit of the poison tree." The phrase was coined by U.S. Supreme Court Justice Felix Frankfurter in 1939. The Justice explained that if information comes from a tainted source, it cannot be admitted into evidence and used as the body of information used to decide a court case.[2] Olaf Hage was careful to apply the same principle and evaluate the lineage of all doctrines he used to determine whether they came from sound sources or not. If a doctrine did not come from a sound source, he did not consider it valid.

You will notice this fact especially when you see how Olaf Hage views the work of author Zacharia Sitchin. While Olaf collected eleven of Sitchin's books, he never considered them works of valid science or history. He was frustrated by the lack of specific footnotes in Sitchin's books. While he had extensive bibliographies at the end of his books and book references at the end of some paragraphs, the lack of page numbers was frustrating. Some of the books Mr. Sitchin referred to were ponderous tomes of several hundred pages. No scholar is going to go through several hundred pages in a source book to verify what passage Sitchin was referring to. Therefore, to be fair to Mr. Sitchin, if his books portrayed an idea, Olaf was careful to give him credit for the idea, but he would also look for other sources that would support or reject Mr. Sitchin's ideas. If he could not find additional support for one of Mr. Sitchin's ideas, Olaf would admit that fact and not give that idea much validity.

Therefore, piecing together the end-times has been an exercise in evaluating the validity of the writings of many of the people. Throughout the Bible, references are made to events that happened, but the Bible does not elaborate about those events. For example, in Genesis 6:4:

> There were giants in the earth in those days; and
> also after that, when the sons of Elohim (gods) came
> in unto the daughters of men, and they bare children

_____
[2]*Nardone v. U.S.*, 308 U.S. 338, citing *Silverthorne Lumber Co., Inc., et. al.* U.S., 251 U.S. 385, 1920

to them, the same became mighty men which were
of old, men of renown. (Genesis 6:4)

The phrase "mighty men which were of old, men of renown" is not clarified in any way. In fact, the author is writing this assuming that the readers of the Book of Genesis would already know what was being discussed. For many years, the readers did know because they were aware of the ancient mythology associated with that phrase. Unfortunately, much of this background information has been obscured by the modern bias of many ministers, pastors, and their academic professors.

Much of this information was "lost" when the "Higher Critical" Movement started teaching their heretical theories in the 18th century in the churches of Central Europe... this new class of "scholars" dismissed the supernatural stories of the Bible as "vain imaginations." Dr. Rudolf Bultmann spent much of his life "demythologizing the Bible." Before long, most Christians attending mainline churches did not believe in *Nephilim*, any kind of giants, or anything else supernatural in the Bible or in any other ancient writings. This cynical academic movement literally divorced people from their history. It also drained people from their faith in Christianity.

Fast forward to the 21st century and now, some people are rediscovering their Christian Faith in that they started to believe that the stories of the Bible were true as written. This true Faith had always been around, but for several centuries, the mainstream church organizations started incorporating the tenets of the "Higher Critical" Movement, which served to undermine that Faith. That Faith started to rekindle en masse in the late 19th and early 20th century when signs and wonders began to break out in Holiness and Pentecostal meetings. People began to accept the validity of the Bible's stories. Eventually, these groups would form up into their own church organizations and compete with other mainstream church organizations for members. After these new church organizations became "part of the landscape," new movements would "come on the scene," bringing more and more people into the historic Christian Faith. As they entered this historic Faith, they began to question much of what the "Higher Critical" Movement adherents had been teaching for a long time.

Those that questioned the old "Higher Critical" teachings began to realize that the Bible "said what it meant and meant what it said." Before long, people began to realize that the Bible spoke considerably about the end times. When this understanding hit the mainstream in the early 1970s, many of the first theories promoted by so-called Bible Teachers proved to be simplistic and were often "fruit of a poisoned tree." When people started checking out these theories, they found that many of these teachings from these Bible Teachers did not begin to consider all the facts presented in the Bible. One of the most overlooked facts of the bible is the Shofar (Trumpet) Judgments resulting from the Seventh Seal in Revelation 8. Each of the seven Shofar Judgments is a cataclysm that befalls the earth. Olaf Hage realized this when he saw the third Shofar Judgment in the Book of Revelation:

> And the third angel sounded, and there fell a great star from heaven, burning as it were a lamp, and it fell upon the third part of the rivers, and upon the fountains of waters; And the name of the star is called Wormwood: and the third part of the waters became wormwood; and many men died of the waters, because they were made bitter. (Revelation 8:10-11)

When Olaf compared this passage to some of the ancient history he had started reading when he was young, he saw a parallel between the star named Wormwood and a star the ancients called Typhon. Every time it visited the earth, it faced a cataclysm.

The most notable cataclysm was the flood in Noah's time. The Book of Genesis considered it a judgment because of the wickedness of the peoples living at the time. (Genesis 6:3) However, it was not the only one. According to Rabbi Louis Ginzberg (a long-time Jewish Theological Seminary professor who was granted an honorary doctorate by Harvard University for his outstanding scholarship and research, but was disdained by much of the Orthodox Jewish Community for his insistence upon understanding the history behind the Jewish writings):

> Nor is this world inhabited by man the first of things earthly created by God. He made several worlds before ours, but He destroyed them all, because He was pleased with none until He created ours. (Ginzberg 1909, 1956, 1994, vol. 1, p. 5)

Rabbi Ginzberg, in his long academic career had studied most of the ancient writings of Jewish literature. He understood that other earth civilizations had existed before our Heavenly Father had created this world. Even the Greek Philosopher Plato wrote about the destruction of Atlantis. The ancient Egyptians also wrote about early earth civilizations that were destroyed before the Book of Genesis. Much of their history was recorded by the famous historian Mantheo. The idea that there was only one civilization on the earth is the theory of Hydraulic Engineer Dr. Henry M. Morris. Olaf Hage, in many of his writings, has evaluated his theories and discredited most them with objective science.

When Olaf Hage further studied the history of the earth, he noted that a heavenly body known in ancient times as Typhon appeared every 5,125 years and wreaked havoc on the earth. He further took issue with the 7,000-year chronology of Dr. Morris, showing that our civilization is much older than what Dr. Morris would have you to believe. Olaf Hage believed initially that Typhon is really two stars that visit the earth. He further determined that our civilization has seen Typhon four times and further, sees that the Bible and other ancient literature have given us warnings about the existence of Typhon. That is how we know Typhon comes in two parts. This book is an enumeration of these warnings, telling us what the ancient world had experienced and what we can expect to experience when Typhon visits us.

The harsh reality is that when Typhon comes around again, it will change how we see the world. It will also allow people to come into power that already understand what Typhon will do and will have the only "plan" to restore our world back to normalcy. Since this cataclysm will be traumatic and a lot of people will die, the remaining people will be more likely to follow those who appear to know what they are doing. However, the Bible warns us that this government will not be godly. They will be like Rome, demanding that we worship them. However, even they know that the second part of Typhon is coming and that it has the capacity to destroy the world once and for all.

While Olaf Hage may no longer be with us, his words of warning have survived him. Every believer in our Heavenly Father needs to read this book and take Olaf's words to heart. The survival of your very soul is at stake.

Chapter 1

Typhon, Reaper of Worlds

"The Axe is laid at the root of the Trees."
John the Baptist (Matthew 3:10, Luke 3:9)

We are by no means the first civilization to carve our mark upon the naked belly of mother earth. The Earth's most ancient residents long ago left the planet covered with mysterious artifacts:

- Megalithic stone circles and pyramids aligned with the heavens, laboriously erected on nearly every continent.
- Hidden caves with colorful, but partially-deciphered artwork.
- Stone sculptures of giant African heads... in Mexico, where they ought not be.
- Oddly similar rectilinear pillars of carved granite figures in lands from Turkey to the bare isles of Scotland... echoed by the strange carved granite statues on Easter Island in the midst of the Pacific Ocean...

All these things appear intended to last for thousands of years, even to the end of this age... In other words, they were built to survive at least until our time.

They also left us writings, when we have not burned them and have retained or acquired the ability to read them: Bizarre legends, incredible histories, scary poems, and most especially, terrifying prophecies.

Perhaps the authors were warning us of the cycle of ice ages. Our scientists tell us the ice will return some day. To allay our fears, our leaders assure everyone an ice age is not a sudden event, but a long, slow, supposedly gradual deterioration over many decades and centuries. Few every bother to check the ice core data, which tells a very different story.

And so, civilization hits the snooze alarm, rolls over, and goes back to sleep...

Yet, after a brief lapse of time, the alarm keeps going off. Something ominous keeps nagging at us, urging us to wake up before it is too late.

What if all these cryptic stones and writings were about something that brings catastrophic destruction, some cosmic object so large that not even our most advanced technology could halt it or divert it?

What if the only possible response is to get out of its way?

The ancients had many names for what is coming. John the Baptist spoke of an "Axe" that is to topple all the trees (Matthew 3:10, Luke 3:9), presumably even the great Cosmic Tree, an ancient image for the tree-like swirl of stars about the Pole Star. To make this discussion easier to follow, I have arbitrarily chosen to use one of those ancient names for this object that topples the poles: Typhon. (By the way, this is NOT Zechariah Sitchin's "Nibiru" object.)

To the ancients, Typhon was a gigantic monster who filled the sky. Typhon so terrified the world, it left haunting footprints upon language: Its tilt of the pole gave Hebrew its word for "North:" *Tsephon* (Strong's Concordance #6828, צָפוֹן). We inherit words for violent storms like Typhoon, and for diseases like typhus and typhoid, for ominous heavenly monsters like Thuban (the brightest star of Draco, the Dragon), a North Star of the ancients, for conquerors like King Typhon, and for his city, Thebes (an ancient capital of Egypt), in whose time Typhon was said to have devastated the earth. The name Typhon was inverted to Phaeton, a god who burned up the world. This inverted form also produced the deadly Python and its Pythian games, a cyclical celebration intended to count time until Typhon's return.

And yes, it will return. No one doubted this. It would return because it has always returned to destroy the earth. And it has returned on schedule. It has a cycle...

The ancients, unfortunately, had lost track of the exact cycle of Typhon. They made a valiant effort to rediscover it... but fell short. Still, their efforts left us a number of clues that we can use to produce a reasonable solution, as I shall show.

Meanwhile, our scientists, although stubbornly refusing to heed the warnings of the ancients (at least publicly) have busied themselves making minute observations of the earth's soil and sedimentation. They have catalogued its tiniest little fossils, even the microscopic bacteria in old rocks. They have published mountains of data that describe, in endless columns and rows of numbers, the precise quantities and measurements of everything.

Our astronomers have been no less diligent. They labor nights, calculating the cosmos they can see, while speculating about the things they cannot. Galaxies beyond seeing have been inferred. Even never-to-be-observed other universes have been proudly declared, as if the mere act of officially imagining them could make them real.

In spite of all of this attention to detail, or perhaps because of it, they seem to have failed to see the obvious... The earth is tilted on its axis. And all the other planets and moons of our solar system also incline in a variety of ways, but usually not so much as the earth.

Why is this important? It takes tremendous outside force to move the Earth's axis. So, if the earth has not had an outside force disturb it, then we have a mystery: What accounts for its axis tilt, not to mention the odd tilts of the other residents of our solar system?

It is known that there is a 41,000-year cycle to the earth's tilting. It "bobs" up and down rhythmically, supposedly moving up and down only about two degrees.

We are told we just happen to be in the midst of that range. So, it can, in theory, only incline a little more or less than it now does. Not to worry... go back to sleep...

But what if that "bobbing up and down" is the dying echo of a more traumatic kind of disturbance in our past? Think of a ping pong ball bouncing until it dribbles to a dead stop. Near the end of its bouncing, the ball is barely a fraction of an inch high off the table. If you had not seen it bouncing almost to the ceiling at the outset, you would never have guessed you were watching the final moments of such a giant initial bounce.

Who is to say that the earth's little squiggles of axial bounce might not be the end result of some horrible past event? When might this have occurred? Perhaps the answer lies in solving another puzzle: Why should there be a 41,000-year cycle for earth's bouncing axis tilt?

Consider this: Why shouldn't the cycle of the axis tilt instead match the supposed 26,000-year precession of the equinoxes, with which it ought to be related, since both work together to determine which star in the heavens will be the North Star?

Yet, there is no known relationship between these two cycles. Whatever cosmic clock rules each of them, it would appear it cannot be the same one. Something unknown is causing that 41,000-year wobbling of our axis...

But would it not trouble you to discover that the mysterious Mayan "calendar" cycle is almost exactly one-eighth of 41,000-years? (Actually, it's one-eighth of 41,003 years.)

The Maya did not assert that the end of this cycle will be peaceful, but rather, that the end of each age is a time of destruction. They expected a rain of meteorites ("gods descending on silken ropes"), ending with planet-wrenching super-quakes and floods and it has just been reported a new-found inscription says meteors will usher in the end of this Mayan age). (*Mysteries of History* 2010) (Velikovsky 1950, 1967, pp. 46-52)

It requires immense energy to tilt our planet. The earth has a huge inertial mass. Either something very big hit us (killing nearly everything), or else something enormous came terrifyingly close in order to have changed the tilt of our axis.

Introduction

There is "nicer" theory, if you prefer to reach for the snooze button of your "smart" cell phone or clock radio. You can pretend that an unknown subtle gravitational tug from something out there, presumably Jupiter (which gets blamed for most things in the solar system that our astronomers cannot explain) may somehow be nudging our axis up and down in a 41,000-year cycle.

But wait. If a tiny, distant nudge takes thousands of years to move the axis in one direction, why should it then reverse itself, and nudge our axis right back to where it was in the first place? And why should this cycle be 41,000 years long? Does Jupiter have the ability to reverse its gravitational effects twice during these 41,000-year cycles?

And keep in mind that, while the North Pole is getting pulled one way, the South Pole is being pushed in the opposite direction at the same time. Quite a trick, even for the gravity of mighty Jupiter.

No, it takes more than Jupiter's massive gravity to accomplish this feat. It takes a bias to distinguish between north and south poles. That bias, of course, is magnetism.

Only an incredibly strong magnetic field from another planet passing very near earth can operate selectively upon our two poles, attracting one while repelling the other.

This does not mean, however, that this other object is not tilted itself. Quite the contrary. It must be tilted dramatically to focus its magnetic field on us in this manner. And if it is tilted so far as to be virtually lying sideways, we may infer that it has itself had a violent encounter with an even larger planet, one big enough to leave our 41,000-year visitor wounded and fallen on its side.

This is a major clue: Such an event is known to have occurred a few million years ago when an object at least as large as earth struck the planet Uranus. Orbiting beyond Saturn, the frozen gas-giant Uranus is fifteen times as massive as the earth. But, like a great Spanish Galleon felled by a heavy cannon-ball, Uranus was toppled onto its side. Unable to right itself, Uranus has become a derelict adrift on the dark sea of night.

So, we know something very big and very dangerous is out there. And its orbit is like that of an immense comet, swinging far out beyond Pluto into the vast Kuiper Belt of ice-moons, then plunging back into the realm of planets, where it wounded Uranus. Its effects are also discerned by the scattering and puffing up of the Kuiper Belt's orbits. Only a very large object, a planet-sized mass, gravitationally jostling these other objects over millions of years, could have disturbed the Kuiper Belt in this manner.

The Maya did not predict the precise day this visitor would come by, but simply said we should now begin to be seeing a lot more things falling from the sky as the cycle of doom comes back to the earth. And indeed, that is exactly what is happening.

Glowing meteoric fireballs are falling from the sky. The number is exponentially growing from year to year, especially at the solstices, the times when the earth's axis is tilted directly toward or directly away from the sun...

It was in 2005 that the number of these bright red and orange fireballs, normally a rare occurrence, began to escalate. By 2009, not only had the annual count doubled, but astronomers were becoming increasingly concerned that sizable impacts were happening on the nearby planets, Mercury, Venus, Mars and Jupiter. The first of these events had been the now-famous Shoemaker-Levy comet of July 16, 1994. Prior to that, astronomers had considered the risk of a modern-day comet impact "extremely remote."

But, in the following three years, two more spectacular comets, Hyakutake and Hale-Bopp, suddenly appeared on orbits lasting 4,000 and 20,000 years, that is, on about the same scale as the Mayan calendar's 5,125-year cycle. Several additional bright comets were seen in the following decade. It seems that a large number of comets have been disturbed by something massive moving through the outer solar system.

Why had such a large object not been detected by our telescopes?

It was becoming more and more obvious that whatever was out there was so dark it would be recognized only by the black void it created when it blocked out the stars in the heavens beyond it. Moreover, the cloud of comets and collision debris that it dragged along with it seemed to be acting like a curtain of cosmic camouflage.

There was one proposed new technology capable of seeing through the debris: An orbiting infrared telescope called "WISE" (Wide-Field Infrared Spectrometry Explorer). WISE made two "mappings" of the sky to enable it to calculate an object's motion to see if it posed a threat to earth. Clearly, finding such threats was a major goal of its mission.

President Obama had promised he would have an "open administration." But after being briefed by top intelligence agencies, he reversed his position. His administration became the most secretive in the nation's history, classifying twice as many documents as all previous U.S. Presidents  combined; this was ten times the number of documents classified at the start of the Clinton administration. (*I.D. Magazine* 2011).

In early 2009, two small comets hit the earth, striking the Pacific Ocean. Then, shortly after Mr. Obama was sworn in, the Air Force abruptly halted its satellite data-feed to astronomers that showed large objects from space entering earth's atmosphere. Not only had objects struck other planets, and the earth, but a primary source for the scientific community's information about incoming cosmic debris had suddenly been classified.

By the time the data-feed was restored in early 2010, the WISE satellite had been launched. For six months, WISE mapped the whole sky. In the summer of 2010, the first report contained a real surprise: There were far more dark objects in the solar system than anyone had ever imagined. We are literally engulfed in a sea of rubble. It is clear that the history of the solar system has been extremely violent.

Even though WISE had not yet made its second sweep of the heavens, a press release declared the earth was in no danger from any of these tens of thousands of newly-discovered dark bodies. Six months later, many thousands of additional objects were announced. Again, without revealing the actual data, it was claimed they posed no threat. Yet, from where came all these thousands of additional new objects in just six months?

The solar system being littered with so much unforeseen new debris supported the Mayan warning that we would be bombarded by an influx of meteors about now. But what could be holding this huge cloud of debris together on a long 5,125-year orbit?

Could there really be a gigantic dark object embedded inside the cloud? Did this dark intruder strike Uranus 3.2-million years ago, just as ice ages began on the earth? Is this dark object the monstrous Typhon of the ancients? How close might it come, and how much damage could it do? Most important, when will it arrive?

The Hopi have a prophecy of two great cosmic bodies at the end of the age, one red and one blue. The Bible, however, has far more to say about the heavenly signs that are to usher in the last days, including these red and blue harbingers of the End.

The prophecies, ancient stones, old legends, and new scientific discoveries are in excellent agreement. To discover how these things, align, we begin our investigation with the ice ages and emerging evidence they begin, not slowly, but literally overnight...

Immanuel Velikovsky wrote about the planetary encounters in our distant past. Typhon is in our immanent future...

Chapter Bibliography

I.D. Magazine. 2011. April: 15.

U.S. News and World Report. 2010. "Mysteries of History." Special Collector's Edition: 30-34.

Velikovsky, Immanuel. 1950, 1967. *Worlds in Collision.* New York: Dell Books.

Chapter 2

Solar Deluges

Death is coming. Not the death we face daily. But a threat of planetary death, "the end of all flesh" of which the Bible warns.[1]

Make no mistake, the whole purpose of the Bible is to save us, and not just humanity, but all the living creatures God has made. The ark of Noah was designed by God to rescue even the "unclean" creatures (Genesis 6:14-22, 7:8). God chose to save even scorpions and rats. And, we know that humanity is more precious to God. (Matthew 6:26, 10:29-31)

Speaking specifically of the coming destruction at the end of this age, Peter said, "The Lord... is long-suffering (patient) toward us, not desiring that any should perish, but that all (persons) should come to repentance." (II Peter 3:9)

So, God's goal is to save, not destroy, humanity. But for that very reason, God must remove those whose acts threaten to annihilate us all (II Thessalonians 1:6-9, Revelation 11:18, 16:6).

[1] See Genesis 6:13, Isaiah 13:6-13, 24:1-6, 34:1-17, Matthew 24:21-22, 37-39, Mark 13:17-20, Luke 21:25-26, II Peter 3:2-13, Revelation 21:1, *etc.*

Genesis records that "the whole earth was filled with violence" because of a mutation or contamination of the genes of not only most human beings, but also of most of the animals (Genesis 6:4, 11-13). Fallen invaders and their *Nephilim* (giant) offspring had "corrupted" both man and beast with irreversible and deadly mutations (Genesis 6:12).

Genesis is being coy. The author alludes to, but shies away from, one big reason God is so "grieved" by this "corruption of all flesh:"[2] Cannibalism!

Other ancient tales of that time are more explicit. They tell us that giants, or "Titans," were eating everyone, even each other. Their genetic code had become like a cancer: They never stopped growing, and so their need for food grew exponentially. Some of these giants exceeded 36 feet tall. (Quayle 2002, pp. 58-80, 113,181-251) (Ethiopic *Book of Enoch* chapter 7:11-14)

Stephen Quayle's documentation of the history of known giants includes many photographs of living men and women in recent time who were from eight to nine feet tall (or taller) (Note that these are also the heights Jewish tradition gives for Cain (nine feet) and his wife (eight feet), who were both reckoned giants). In his book, there is also a photograph of a fossilized giant human twelve feet two inches tall. (Quayle 2002, facing page 256)

Stephen Quayle also found references to unearthing human skeletons of enormous size in historic times. For example, skeletons of 34 and 36 feet were found near Athens in Greece. And, in Bohemia are kept the leg-bones of a 26-foot human found in *A.D.* 758. In Sicily, in 1516, a giant over 30 feet tall was found near Mezarino, and others were found there of about the same height in 1548 and 1550. And a 25 foot 10-inch giant's bones were said to have been dug up in France in the year 1613. A nine-foot giant was found in Crittenden, Arizona in 1891. There are dozens more examples in his book. (Quayle 2002, p. 258*ff*)

When gigantism became an uncontrollable cancer-like gene, <u>growth did not</u> stop in adulthood, but continued endlessly. The

[2] Genesis 6:4-7, 12

Ethiopic *Book of Enoch* (Chapter 7:11-14) claimed some giants grew to heights of "hundreds of cubits" (= over 340 feet tall).

Earth could not sustain the giants' unrestrained appetites. They ate everything. It was as if gigantic malignant cancers were devouring all living tissue. Worse, they had corrupted all animal life. At the very least, other creatures were being forced to eat each other to survive. The whole world had descended into a devouring madness.

If God had not come to our rescue, there would have been no flesh saved alive. Mankind and the animals had fallen into bloodthirsty depravity on a global scale (*cf.* The Ethiopic *Book of Enoch* 7:11-14). A planet-wide crisis of this kind cannot be solved without some kind of planet-wide solution.

Most ancient cultures that had legends about giants also had myths about how God (or "the gods") were forced to bring about global flooding or other planet-wide upheavals. Donald Patten, Ignatius Donnelly, Robert W. Felix, and many others who have written about catastrophism in myth and legend over the last 250 years. (Quayle 2002), (Velikovsky 1950, 1967, pp. 46-52)

There are several alternative ways God could intervene:

He could let all life be devoured by the giants and wait for them to starve to death. He could then start over again. Or, He could let a cosmic impact obliterate all life. Both "solutions" would fail to preserve any of God's original creatures, and both would break His Messianic promise. (Genesis 3:15) So, both of these options were unacceptable.

The third alternative, which was the only one that could preserve any of mankind and thereby keep God's promise, was to use the "surgical" tool of climate change:

But most of the time, climate reshapes the earth slowly: Seas recede, Mountains erode. The coastlines gradually break-up and deserts expand into forests. To God, these changes may seem to pass rather rapidly. He sees the earth as a fluid in constant motion.

Of course, God's perspective is not ours. We feel that the planet beneath our feet is a hard, unchanging ball of rock. A big, solid planet earth would be comforting.

But, alas, the earth is not solid. It's liquid.

And climate can affect liquids, even in the core of the earth, which is mostly iron heated into what is essentially a liquid state. The mantle, which occasionally erupts with volcanic lava, is clearly also in liquid form. Even the "hard" crust, which is precariously thin, is peppered with vast pockets of lava, oil, gas, and water. And three-fourths of the surface of the earth is water, most of it in the great oceans.

That leaves but a tiny part of our planet that can be described as habitable earth. Nearly all of that is in the seven continents, which are relatively slim rafts of cooled lava floating in the great sea of molten rock beneath the crust.

Thus, the continents are "frozen" lava, otherwise known as granite. The continents float in a sea of liquid lava, and as floating bodies, they bob up and down.

Climate change can make continents rise and fall, flooding them periodically. However, such floods would have to be very sudden and unexpected to sweep away the giants. We will see that this is exactly what happened in Noah's Flood. But the precise mechanism required, which is rather disturbing, is known as "tectonic de-leveling." It generates mountain-high, globe-sweeping tidal waves.

The climate has several ways to move the continents. First, it can wear down one side while building up another. Tides eat away at coastal beaches, while rivers move silt down to the shore and build deltas. It is a slow, cumbersome process that moves one tiny grain of sand or silt at a time. Hardly a good way to stop giants from eating mankind.

A second way for climate to move continents is by expanding the oceans. Hotter water expands and pushes the continents apart. Again, it is painfully slow.

The third way is by rain. The continents began as cold, hard lava. Water at first ran off, ending back in the sea. But as climate let life prosper, plants took root and slowly ate their way into the granite, making it more porous, capturing moisture, which slowly increased the weight of the continents. That increased mass caused them to press down into the lava below. When climate warms sufficiently and rain ceases, deserts spread, water evaporates, and the continent lightens and gradually rises up again. This process is also excruciatingly slow.

The fourth option is not slow. It comes as a torrential deluge of rain... or ice that does not melt, but rather, continually accumulates, day after day, rapidly piling itself up higher and higher, until it becomes an ice-mountain. At that point it forms a glacial mass that becomes self-reinforcing. Like a runaway nuclear reactor, the ice voraciously feeds its own gigantic glacier. Every drop of humidity that blows over it immediately freezes onto the towering cliffs of white death. Nothing can grow under it. Nothing can breathe under it. Every crevice is filled, every tree crushed, as it slides inevitably over the land, constantly spreading and building itself up, like a cancerous growth that nothing can halt. The ever-growing ice mimics the ever-growing giants who are swept away when it melts.

The weight is almost beyond calculating. At the peak of the recent ice age, a scant 18,000 years ago, North America had a glacial monolith seated on its spine like some corpulent emperor devouring everything before him. This imperial glacier soared up into the cold blue sky some three miles above the surrounding land. It could be seen for hundreds of miles to the south, where the land sloped down to the Gulf of Mexico, much shallower then, waiting for the flood of glacial melt-water.

The mechanism that turns ordinary winters into an instant glaciation has been a mystery. Dozens of theories have been advanced to explain why climates just like ours switch seemingly over-night from warm life-enhancement... to cold glacier-building.

One idea is that every winter is a gamble. If enough snow happens to fall, it will not melt completely in higher elevations. That forms a new base of ice for the next winter and so on. Any given winter could begin a new ice age. Yet, if that were true, ice ages would be more frequent and random. So, by itself that cannot explain ice age cycles.

Another popular explanation is the valve theory. This idea says the ocean currents sometimes pass through shallow seas and narrow passageways that can get shut off like a valve. Unable to take its old route near the continental land mass, the warm ocean current flows out to sea, leaving the continental coastland to cool and begin to build up snow and ice. The problem is, to shut off the valve, you need falling sea-level and that requires prior ice-building on the land. So, the valve theory might contribute to rapid ice accumulation, but it probably could not trigger the initial build-up.

Another idea is that ice itself does the trick. In this theory, ice builds up at the poles, destabilizing the earth's rotation. The planet then tilts, which plunges previously warm continents into arctic regions and vice-versa. Prof. Charles Hapgood popularized this theory in his book, *The Path of the Pole*. His  mechanism looks effective at first, but the sheer energy needed to change the poles as he described would seem way beyond the capabilities of polar glacial build-up. (Hapgood 1958, 1970, 1999)

As we can see, creation of an ice age is not as easy as we might assume. Indeed, one of the biggest puzzles about ice ages comes from the large-scale history of the earth. The standard model says earth began as a molten ball, but the sun was then 30% cooler. So, the earth may have frozen briefly at its surface (between meteorite bombardments).

After our solar system orbited the galaxy for the first time (*c.* 200-million years), a Mars-sized planet collided with the earth, perhaps twice, forming the moon. (Raeburn 2012) As the moon receded from the earth, huge tides were generated, cooling the molten wounds in earth's mantle. These diminished as time passed and the moon moved further away. But tidal friction kept the sea from freezing for a long while.

Eventually, however, the moon had moved out enough that the tidal action could no longer offset the early sun's lack of heat. The earth froze solid.

This ice age was the greatest of all. It lasted roughly three-billion years. Then about 600-million years ago, a mysterious cosmic bombardment shattered the sea's blanket of ice. That allowed life to ascend from the deep and get up into view of sunlight, which had by that time grown warmer. The last vestiges of that bombardment still continue.

With sunlight now penetrating the sea, the ultra-violet radiation began to mutate the tender tissues of the earth's jelly-fish-like sea creatures. Their natural defense was to use the light to fix calcium, creating shells and bones and teeth and horns and claws and various other hard structures. These made it possible for life to emerge from the sea, out into the air, where a creature needs a solid skeleton to stand and walk.

Ice returned periodically after that. Some scientists think cosmic bombardments every 30-to-32-million years, on average, may account for great extinctions; afterwards the sky was darkened with dust and debris. So, the temperature fell, and ice built up. And, that made sea-levels plunge. (Gribbin and Gribbin 1996, pp. 36-40, 167-196)

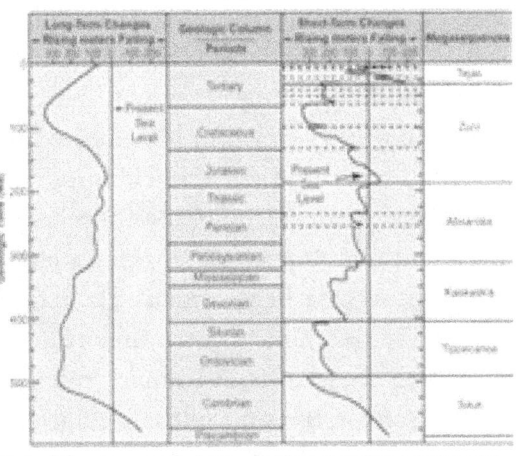

There has been more than a dozen of these brutal sea-level plunges over the past 600-million years. (Morris and Johnson 2012) Every such plunge implies a very rapid build-up of ice on the continents. Afterwards, the earth slowly warmed again, and sea-level gradually recovered, only to be struck by yet another instant ice-age before the planet could return to its previous high temperature. (Morris and Johnson 2012, p. 13 chart citing Sloss, Vail and Mitchim.) For over 300-million or so years, the earth was repeatedly battered, forcing sea-levels progressively lower, until the ice seems to have built up to a global level not seen since the great ice epoch ended 600-million years ago.

Nevertheless, it appears that in-between these cosmic impacts the world remained fairly stable for long ages. The most famous of these periods was dominated by the great dinosaurs. Dinosaurs are now recognized as a necessary transition between clod-blooded reptiles and warm-blooded mammals. But anyone who has seen a *Jurassic Park* movie realizes that mankind could never have prospered if we had lived during that era. It was necessary for the dinosaurs to be wiped out so that mammals like us could proliferate. It is easy to see why God had to cut short their reign if we were ever to come on the scene.

With the dinosaurs gone, the earth's mammals grew dominant. Then, as if on a strict schedule, around 35-million years ago, the earth again was struck by a large object, whose crater formed Chesapeake Bay. This ushered in a long dry ice-age and sea-level plummeted. (Gribbin and Gribbin 1996, pp. 82-83) Mammals were forced to become more intelligent in order to stay alive in the arid climate. Again, we can see God's hand quietly at work.

About 2.3-million years ago, the climate suddenly changed again. The world seemed to have a new kind of ice age. Ever since, the ice has come and gone much more rapidly. Intervening warm ages are now very short, 10-12,000 years or so. The current warm age is about 11,600 years old, making another bout of ice immanent.

Our warm age began when a flood of melting ice raised global sea-level 375-feet. This was the time Plato understood to have been the sinking of Atlantis. And many researchers (including myself) believe it sank at the time of the Deluge of Noah.

The Bible says that God sees the end from the beginning (Isaiah 46:10). He foresees the ultimate goal from the beginning of the creation. He has mankind in mind when the earth was formed. When it brought forth fish, God foresees animals on the land. When it brought forth animals, God foresees humanity and our destiny.

We are told Christ was "the Lamb slain from the foundation of the earth." (Revelation 13:8). God is patient like a Potter (Isaiah 43). He forms mankind out of the once-lifeless dust of the earth and gradually transforms us into fit vessels for His Spirit. (Genesis 2:5-7)

But God's method of forming the earth and its creatures involves the use of cosmic bodies like the sun, the moon, and other objects that orbit through space. They are His arms and hands (Exodus 6:6, Deuteronomy 9:29, 32:41-42, Psalms 44:3, Isaiah 34:5-6, Ezekiel 21:3-5, etc.). With them, He sculpts the mountains and carves out the seas Yet, having given living creatures free-will, God must occasionally intervene to protect life from total destruction. So, God tells Noah how to build an ark in order to save animals (and humanity) from "the end of all flesh" in the Deluge. (Genesis 6:13-16).

The Bible says God does not change (Malachi 3:6). He is the same yesterday, today, and tomorrow (Hebrews 13:8). So, we can infer that whenever humanity and life are in dire danger, God intervenes. Such a time, Christ said, is coming soon, at the end of this age (Matthew 24:21-22, 37-39, etc.). Jesus even draws attention to this parallel with the Deluge near-extinction by saying, it will be as it was in the days of Noah. (Matthew 24:37-38, Luke 17:26-27)

Although most people assume "evolutionary" science sees geo-history as uniformly gradual and does not believe in recent global human extinctions, modern science now recognizes several recent near-extinctions of mankind. During these "evolutionary bottlenecks," the entire human race was almost obliterated by sudden climate change, namely ice-ages. These die-offs seem never to be caused by global warming, however.

In one case, roughly 47,000 years ago, as few as 5,000 humans survived, this die-off was from c. 60,000 to 40,000 years ago, but closer to 44,000-50,000. (Adam, Eve, and the Genome 2009)

Some scientists believe it was also about this same time that the genetic ancestor ("Adam") of all modern men lived, which implies all the other males of that time left no descendants at all today, another indication of mass-extinction. (*Adam, Eve, and the Genome* 2009, p. 47)

In another case, about 70,000 years ago, less than 2000 people are now thought to have been left alive because of a sudden ice-age onslaught. (*Retracing Our Steps Back to the First Humans* 2009)

Even more dramatic, about 159,000 years ago (±36,000 years), the total human population plunged to a few hundred people, according genetic studies. (Marean 2010)

We should explain what these genetic studies are. Samples are taken from all the branches of modern humanity. Then specific genes are isolated and compared to create genetic "trees" that show how the different genes were derived over time throughout mankind. These DNA trees can be compared with each other and datable information from the archeological record. These comparisons can be rather complex. So, the data is compiled in a computer, where thousands of bits of information can be sorted out. This painstaking process leads to implied (albeit rather uncertain) dates when various genes first emerged and how many individuals with that gene must have been alive at a given time. (*Adam, Eve, and the Genome* 2009, pp. 42-49)

But in one of these massive genetic studies, something earth-shaking emerged from the data. The scientific team found there was a time when there may have been but two survivors, as recently as 23,000 years ago. Hence, mankind has come from just two people who may have lived only c. 23,000 years ago... one man and one woman. This finding led even scientists themselves to begin using references to Genesis to describe it. (Brown 1990, pp. 32, 211, 236)

It is not a little disconcerting that mankind has teetered on the edge of annihilation so many times. But the thing that is truly scary is that, if one looks at the timing of these near-extinctions of mankind, one can discern a rhythm in them.

The common denominator of 23,000 and 47,000 and 70,000 and 159,000 years is 23,000 years. If we look at other known extinctions, we find even more events on this 23,000-year cycle, for example, one in the British Isles 93,000 years ago. (Felix 19097) Mr. Felix came to the same conclusion about the cycle length as I had when I studied it in the 1970s.

In other words, these near-extinctions keep happening, over and over, according to a predictable interval. And given the length of the time interval, the cause of these horrific events must be cosmic, operating on a cycle that takes about 23,000 years. And, accordingly, unless something intervenes, it will happen again.

And given the possibility that one of the most recent calamities was about 23,000 years ago, we can infer that another such time of near-extinction is coming soon. That is exactly what Jesus warned us would happen at the close of this age.

However, there is a problem. Complicating the calculation of when it will happen is evidence for more than one cycle of destruction. That is, there are several repeating series of catastrophe in the human record, and many great investigators have stumbled upon this unsteady ground. They made the mistake of assuming there was only one cycle at work. So, they have tried to explain all mysteries with but one simplistic explanation.

But our earth has been battered by more than one cosmic cycle. The rhythms of destruction form a symphony of several cycles that weave in and out together. Some are quite long, and some are fairly short. The famous "Mayan" countdown of *c.* 5,125 years is an example of a shorter cycle. We shall decipher that mystery later on in this book.

The most scientifically studied cycles are solar. The solar cycles are driven by planetary orbits, primarily Jupiter's orbit. As the planets travel about the sun, they tug and pull on the solar globe, shifting it slightly this way or that. But if all were to mass on one side, their pull could move the sun one solar diameter out of place, 864,000 miles. Usually, the planets only nudge it, making the sun shimmy like a ball of white-hot jelly.

If (as science claims) the sun is a fusion reactor, precariously balanced at its core, these shimmies should disturb solar output. It is thought to take about 10,000 years for these disturbances manifest at the surface of the sun and radiate out to the planets. While these changes percolate up to the surface, they interact, which normally dampens the changes, by averaging out all those gravitational tugs and pulls from the planets.

Unfortunately, the sun accumulates and reinforces a lot of these smaller changes, and they begin to generate cycles of greater and lesser solar activity, periods of general warming or cooling that rise and fall in great waves.

For example, everyone knows about the eleven-year sunspot cycle, but it can actually range between seven years and fourteen years long. The long-term average is currently 11.135 years (over the last century or so). Two of these sunspot cycles run about 22.5 years and are known as the Hale magnetic cycle because the solar magnetic field "flips" in each cycle, forming a pair of positive and negative cycles. The 22.5-year Hale cycle has been linked to a 22-year drought cycle in the United States.

Four sunspot cycles total about 45 years and correspond to cycles of dramatic auroras--times when strong solar winds interact with earth's North Pole. These 45-year waves bracket periods of time when the brightness of the outer planets (Mars, Jupiter, Saturn, Uranus, Neptune, and Pluto) and their temperatures (as measured by space-born infrared sensors) increase or decline. They all rise and fall with the earth's temperature, getting gradually warmer for 45 years and then becoming cooler for 45 years.

The so-called "global warming" hysteria of the 1990s and 2000s was the final phase of the latest 45-year solar warming cycle. All the outer planets had also warmed during the same time-frame (a fact long known by planetary physicists). It is assumed Mercury and Venus are also warming and cooling in sync with the rest of the solar system, but they are too "bright" and close to the sun to risk pointing delicate infrared sensors at them.

Eight 45-year cycles accumulate to yield a 360-year cycle of large climate waves. This is about the time it takes Jupiter to "process" against the galactic background. There are about 360-years from the lowest temperature phase (the "Little Ice Age" of the 1600s) up to the warmest phase (where we are now). Seven of these cycles total c. 2,520 years, a cycle which shows up clearly in the C-14 levels of tree-rings. It may also be the "seven times" the Bible cites (7 x 360 "days," reckoned as a year for each "day") (Leviticus 26:18ff).

Whatever one may believe about C-14 as a dating tool, the mechanism by which this form of radioactive carbon is generated, is not disputed by anyone. The process begins in the upper atmosphere, where nitrogen abounds. Our atmosphere is composed of about 80% nitrogen, which has fifteen protons and fifteen electrons. High-energy cosmic rays from the galaxy bombard this high-altitude nitrogen with such force they can knock off one of these protons and its accompanying electron. This cosmic impact leaves behind an atom with fourteen protons and fourteen electrons. But that is no longer nitrogen; it is radioactive carbon-14. The carbon-14 eventually falls to the earth where it is taken up by plants.

If nothing intervened, this cosmic-ray bombardment produces a constant amount of new carbon-14. But a stream of charged particles from the sun blows outward past the earth. This is the solar wind. When it encounters the galactic cosmic rays, the solar wind's magnetic field pushes them away from the sun and the inner planets.

However, the solar wind is not always the same strength. It fluctuates with all of the solar cycles we have been describing. So, the ability of cosmic rays to generate more carbon-14 is reduced when the solar wind is stronger, but more carbon-14 is generated when the solar wind is weaker. This was proven by C-14 changes in the 11-year cycle.

Because atmospheric C-14 is directly regulated by the strength of the solar wind, this 2,520-year cycle shows that long-term solar fluctuations are quite real. We can map in some detail these 2,520-year cycles of solar activity over thousands of years. Ten of these 2,520-year cycles are almost a precessional cycle: 25,200 years. The total falls short of the ideal Platonic precessional cycle of 25,920 years by 2 x 360 years, that is, by one full climate fluctuation from one "Little Ice Age" to another. 36,720-year climate waves are in a 25,920-year precessional cycle, or three cycles per zodiacal "age." Each 720-year cycle matches one of 36 astrological "decans" or ten-degree segments of the zodiac, the 36 traditional "signs" not in the zodiacal band where planets orbit. Knowledge of 720-year climate cycles may have been encoded by the ancients in these 36 "extra" signs.

So, the earth is on a long-term roller-coaster of warming and cooling. Right now, we are at the hot top of the climate roller-coaster, and from here on we will be going downhill, getting progressively colder for the next 360 years.

Starting to get that "sinking" feeling?

Well, before you relax and begin thinking that 360 years is a long way off, you need to know that every so often during one of these cycles, the earth's cooling climate reaches a sudden break-point where the planet plunges precipitously into full glacial conditions. And this has happened a lot.

There is a long and scary record of these sudden ice-ages in the geologic record. Although climatic warming tends to be gradual, the ice ices come brutally fast. And it is this quick-freeze that is associated with rapid die-offs of humanity. The ice seems to fall instantly from heaven. And this is exactly the kind of event that the Bible is warning will happen at the end of this age. (Job 38:22-23, Revelation 6:15-17)

We had better find out what causes this ice event, for we may be the generation that will be forced to confront it.

One clue is that it seems to be a deluge of ice, like the deluge of rain that Noah survived. In fact, some of the ancient "deluge" stories refer not to rain at all, but to snow and hailstones. Examples of this include the Norse legends and the traditions of the Polynesians. The one common feature in every deluge story or ice-event is the sudden massive quantity of precipitation. The amount of moisture falling from the sky must be horrendous, regardless of whether it is rain or snow or hailstones or ice.

Where does all this moisture come from? Obviously, it falls from the sky. But how does it get up there? Cold cannot cause an ice age, because cold causes evaporation into the atmosphere to decline, not increase.

Water in the atmosphere comes from either cosmic ice or evaporation. But space probes to comets have greatly reduced the estimates of the amount of ice in the typical comet. Unless we were to be hit by a truly huge comet, it would be difficult to generate the kind of rainfall described in these deluge stories. But a comet that large would impact the earth with an energy level that would literally convert the atmosphere into a fire-storm, not a deluge. (Gribbin and Gribbin 1996, pp. 33-38)

That leaves us with evaporation. But here too we have a problem. In theory, the sun must be the mechanism by which evaporation would lift sufficient moisture into the sky to supply enough rain for a deluge. But how can long, slow cycles of solar activity produce a sudden catastrophic deluge event? How can the sun evaporate up such large quantities of moisture needed to create a global flood, without it raining out as usual?

The answer involves the color of the sun, or more correctly, the amount of the sun's energy that is generated in the various parts of the solar spectrum, not all of which are visible light. In fact, most of the solar spectrum is invisible. The light we see is actually a small part of the solar energy output. Ultraviolet light, for example, is invisible, yet without it, photosynthesis cannot occur, and plants would not grow.

Changes in solar activity are accompanied by changes in the solar spectrum. So those cycles we discussed earlier are also cycles of solar color change. When you think of the color of the sun as you were growing up, if you are old enough, you will remember it as being more golden than it now appears. Today it looks whiter or bluish. That is not as subjective as you might be thinking. The color does fluctuate.

Why is the color of sunlight important to evaporation rates? Not all parts of the solar spectrum can get past the atmosphere with the same degree of efficiency. Ultraviolet light has fairly little trouble penetrating the atmosphere, but down at the opposite end of the spectrum, infrared radiation has a very difficult time passing though the atmosphere.

And if light cannot get through the atmosphere, it cannot evaporate water back up into that atmosphere. This means that when the solar spectrum is primarily concentrated up toward the bluish ultraviolet, it has an easy time piercing through the atmosphere and reaching the seas, where it can heat up the water enough to evaporate it. But if the spectrum of sunlight were to be concentrated in the redder end, which is far less efficient in getting through the air, then far less evaporation occurs.

Now we can see why fluctuations in solar activity during the sunspot cycles are able to correlate so well with drought cycles in the American southwest, which depends heavily upon the rate at which the sun evaporates moisture out of the Pacific Ocean.

As the sun's spectral output shifts up and down, the total actual energy output of the sun changes very little. By contrast, a small shift in solar gross energy output can result in a large shift in the solar spectral output. And a large change in the spectrum can play havoc with global rainfall.

Now we have a mechanism which can indeed generate a deluge under the right conditions. But there are several steps in the process:

First, there has to be an extended solar ultraviolet cycle. The sun needs to keep pouring out bluish light at an intensive level for several years. That will evaporate up enough water to saturate the atmosphere with a potential deluge of water or ice.

But what prevents this water from raining back down again the way it usually does? What holds back the rain long enough for a deluge of water to accumulate?

Ironically, a massive global deluge requires a massive global drought. Nothing causes rain more effectively than dust. Conversely, a lack of dust would result in a drought.

When moisture rises up into the atmosphere, it collects around tiny grains of dust and then freezes. As more moisture contacts this tiny ice crystal, it also freezes. When we get enough ice to form, the weight causes it to fall out of the cloud. Generally, if the air below is warm enough, it melts, causing rain. If it does not melt, we get snow or ice.

But without the dust, we do not get the rain at all.

This fact has been driven home to me personally here in Maine. When the snow builds up, I open the window that looks out onto the roof and scoop snow out so that it won't melt back into the window and drip inside. The snow I remove is dumped into the bathtub to melt. But when it does, it always leaves behind a layer of sooty dust particles from inside the snow. Somewhere up in the atmosphere, this dust had formed the nuclei to create snow and ice in the clouds.

So where do we get the dust from? Some of it can come from drought itself, if a dust storm whips the particulate up into the atmosphere. But this is a very ineffective method. Dust bowls can come and go without causing rain. (Bartholet 2012)

Another dust source is volcanism, but it is also intermittent. On the other hand, volcanoes can inject ash and dust into high altitudes. So, their dust is readily available for rain formation. If volcanism stops for a time, that can reduce rainfall compared to times when volcanoes are more active. If the weight of ice on the continents changes, that might well affect how much volcanism takes place.

But there is a third source: Cosmic dust.

The earth is continually plowing through a vast sea of dust that floats between the sun and the asteroid belt. This cosmic dust-belt is actually visible to the naked eye just after sunset on a clear evening. It is known as the Zodiacal Light and stretches upward at an angle, to the left of the setting sun in a long narrow band. It is believed to be the residue of thousands of comets that have disintegrated during their passages near the sun.

In 2007, Comet Holmes exploded, while passing through the asteroid belt on its way out of the solar system. It may be assumed that the comet hit something and burst asunder. So, it would have injected a lot of dust into the asteroid belt, not around the earth.

As the earth and other planets pass through cosmic dust, it is steadily gathered up and removed from the belt. So, the belt in the earth's path erodes constantly. Even with the comets breaking up from time to time, the amount of dust the earth sweeps up every year must more than offset what these comets contribute within our orbit, on average.

After many years, the dust should have been mostly absorbed by the planets. Yet, there is enough remaining dust that we can see it with our naked eyes as a visible violet glow that resides distinctly in the plane of the planetary orbits.

That is not what we ought to see. If the Zodiacal Light were the product of the usual influx of disintegrating comets, then it should not be a flat planetary band, but a sphere of dust extending equally in all directions. Typical comets dive down into the solar system from all directions, not just in the plane of the planets. If such comets were the source of cosmic dust, there should be a random distribution in the shape of a sphere.

The narrow band of cosmic dust implies that it comes, not so much from comets that normally pass through the solar system, but from an extremely large source that orbits within the planetary plane. That suggests something connected with the asteroid belt, which is a similar flat band. The appearance is reminiscent of the planet Saturn, which has alternating rings of debris and dust surrounded by moons of varying sizes.

The rings of Saturn are said to be slowly eroding as well. To replenish the rings, an occasional large object has to get close enough to Saturn to break up. Therefore, the dust of the Zodiacal Light must be periodically replenished, apparently by some large cosmic body surrounded by a vast cloud of dust that accompanies it.

If a large dust-bearing object were to pass near the earth's orbit periodically, then we might have a mechanism for creating cyclical changes in global rainfall (or snowfall). An increase in dust would result in a period of elevated rain lasting many centuries. Eventually the dust would be gathered up, and global rainfall would slowly decline again. Toward the end of the object's orbital period, just before its return, global rainfall will have declined to drought levels in many places, as is currently the case.

If the sun became extremely ultraviolet near the end of this cycle, it would cause a lot of evaporation with relatively less rainfall. Under those conditions, a large amount of net evaporated water would accumulate in the atmosphere for several years. A horrific amount of evaporated water could theoretically build up over our heads.

Okay, but how do we get all that water to fall back down?

The heat of the ultraviolet sun can support the moisture overhead as long as the sun continues to pump out its energy in the upper blue portion of its spectrum. But if that should change and the light began to sink down toward the red end, the ability to suspend water in the atmosphere would drop. An infrared sun would be unable to keep the rain from falling.

It would look blood-red for a limited time as the spectrum shifted downward. But then the sun's color would darken, growing brownish, then black. As the Bible puts it, the sun would appear "black as sackcloth of hair," as seen from the earth. And at the same time, the Bible says, the moon would look red as blood (*cf.* Revelation 6:12). The infrared light from the sun would look black when viewed through the earth's atmosphere, but when the same infrared light is reflected off the moon's surface, it would be refracted and be seen as blood-red from the earth.

Now we begin to see what the Bible is describing. In spite of the poetic language, the text is scientifically precise. An infrared sun is exactly what is needed to produce the deluge of hailstones described in Revelation 16. And an infrared sun would look just as Revelation 6 describes.

So we seem to be onto something here. Solar spectrum fluctuations, triggered by long-term planetary cycles that generate disturbances in solar output, can evaporate up huge amounts of water vapor and then dump it catastrophically as a deluge of hail.

What all of this can then yield is an ice-age. The ice builds up tremendously fast. As we noted previously, the ice becomes a self-reinforcing reaction. It rises up into a mountain that gobbles up any moisture that touches it. An entire hurricane of moisture can be devoured in a day, locking it up in a great continental glacier. As the prevailing winds pump water out of the oceans, the glaciers convert the water into ice with astonishing efficiency. The weight of water in the sea basins goes down rapidly, and the weight of ice on the land becomes unbearable.

The turgid motion of continental drift in the underlying molten mantle creates a severe inertial resistance. As the weight escalates, the pressure for the continental slab to move becomes enormous. It grows geometrically until it reaches the breaking-point. That moment comes when the straining continent lurches suddenly in a cataclysm.

The whole continent sinks at one time. The oceans on all sides surge up hundreds of feet and begin racing out in every direction. In some case the tidal waves tower up a mile high. Nothing in their path can survive. The impact force is like having a wall of iron slam across the surface of the sea. When it reaches the shore, any vestige of ancient civilization would have been obliterated and swept far from where it stood.

So, we can get a tidal-wave flood when continents suddenly sink. But the same thing happens as the ice melts. And there was indeed a sudden melt-off 11,600 years ago when a torrent of melt-water filled the Gulf of Mexico. The only surviving fragments of that lost world are far inland and on high ground, such as the sparse remnants that have been found high on the Turkish plateau. Archaeologists there are unearthing cities of heavy stones, and little else. There are signs of a once-extensive civilization that existed about 11,600 years ago, with towns scattered all across a vast region of the Turkish and Syrian highlands. (Mann 2011) (*The First Sacred Place* 2010)

What destroyed that civilization? Only heavy stones, up to ten tons, have survived. Megalithic stone is the one thing a tidal wave has trouble moving.

Geologists have studied the rocky coast of North America on the western shore of the Atlantic Ocean, halfway around the world from Turkey. The northeast American coast had been weighted down with ice until about 11,600 years ago. Today that rocky coast has a ledge of stone that runs from Canada down past New England, fractured here and there, even to the Carolinas. It was the reason for the Industrial Revolution. Rivers tumbling over that ledge, or fall-line, generated the water power that ran the mills of early America. Modern culture might not have been created without that ledge of broken rock.

But that ledge is also a surviving remnant of one of the most stentorian upheavals of recent geological history. It is part of a fracture-line produced when the continent suddenly jolted up out of the molten muck of lava, into which it had been depressed by the great glaciers. The result was a titanic, hemispheric tidal-wave that surged eastward and southward until crashing into Europe and Africa. Nothing in its path survived.

Including Atlantis... Plato's account (in his *Timaeus Dialogue*) records the last days of the islands directly in the path of the tidal wave. The visiting Athenian Greek statesman Solon is told the story of Atlantis by an elderly Egyptian priest of Sais, a city upriver from Alexandria, founded about 250 years after his visit *c.* 580 B.C:

> O Solon!" (The Egyptian priest said,) "There have been, and there will be again, many destructions of mankind arising out of many causes... (This) signifies a declination (close approach) of the bodies moving around the earth and in the heavens (the orbiting planets and comets), and a great conflagration (burning up) of the things upon the earth recurring at long intervals of time... When, on the other hand, the gods purge (cleanse) the earth with a deluge of water...those of you who live in cities are carried... into the sea... then, at the usual period, the stream from heaven descends like a pestilence, and leaves only those of you who are illiterate and uneducated, and thus you (Greeks) have to begin all over again like children, and know nothing of what happened in ancient times... (whereas) all that has been written down of old... is preserved in our temples... for, in the first place, you remember one (most recent) deluge only, whereas there were many of them...because for many generations the survivors of that destruction died and made no sign. For there was a time, Solon, before that greatest Deluge of all, when the city which now is Athens was first in war... The goddess who is the common patron and protector and educator of both our cities (Neith)

founded your city (Athens) a thousand years before ours (in Egypt), ...and then she founded ours, the constitution of which is set down in our sacred registers as 8,000 years old (= c. 8,600 B.C.) ...Our histories tell of a mighty power (Atlantis) which was aggressing wantonly against the whole of Europe and Asia, and to which your city put an end. This power came forth out of the Atlantic Ocean, for in those days the Atlantic was navigable; and there was an island situated in front of the straits which you call the Columns of Hercules. The island was larger than Libya and Asia put together (North Africa and Turkey combined), and was the way to other islands, and from these islands (the Bahamas, Virgin Islands, *etc.*) you might pass through to the whole of the opposite continent which surrounded the true ocean; for this sea which is within the Straits of Hercules (the Mediterranean Sea) is only a harbor, having a narrow entrance (Gibraltar), but that other (the Atlantic Ocean) is a real sea, and the surrounding land (America) may be most truly called a continent. Now, on the island there was (the capital of) a great and awesome empire, which ruled over the whole island and several others, as well as over parts of the (American) continent; and besides these, they subjected the parts of Libya (North Africa) within the columns of Hercules (Straits of Gibraltar) as far as Egypt, and of Europe as far as Tyrrhenia (Italy). The vast power thus consolidated into one (empire), endeavored to subdue at one blow our country and yours, and the whole land within the straits. Solon, your country...triumphed over the invaders, and preserved from slavery those who were not yet subjugated, and freely liberated all those who dwelt within the (Straits of Gibraltar = the Mediterranean

Sea). But afterward (*c.* 9,600 B.C.) there occurred violent earthquakes and floods, and in a single day and night of rain (no ordinary storm, but "that greatest Deluge of all" which fell upon both Atlantis and Athens), all your warlike men together sank into the earth, and the island of Atlantis in like manner disappeared, and was sunk beneath the sea... Many great deluges have taken place in the 9,000 years (since these events)." (Plato 2015)

Note that Atlantis was plagued by violent earthquakes and floods for some time prior to the sinking, and that the whole Greek military "sank into the earth" on the very same night that Atlantis sank. Moreover, all those who lived near the Aegean Sea drowned. This was a deluge of rain that covered all their known world, from America to Greece and Egypt. We must infer that Atlantis sank on the first day of the Deluge, having already been undermined by unusually violent earthquakes and violent floods.

The term "violent earthquakes and floods" may refer to a tectonic de-leveling of North America as the ice-sheets melted in the days leading up to the Deluge. A "violent... flood" suggests a tidal-wave, not just rising waters. Tsunamis are caused by unusually violent earthquakes. If the Egyptians had recorded these quakes striking the far side of the Atlantic Ocean, they had to be of stupendous magnitude. And the floods, the priest said, reached as far as Greece.

The text does not literally explain that the Atlantic Ocean, that is, the "Ocean of Atlantis," had abruptly swept into the Mediterranean Sea and had destroyed Greece and Egypt. But something connected to the "violent" flooding of Atlantis had also drowned the Aegean Sea, and it takes no great insight to see the connection was the tidal-wave caused by the North American continent uplifting, which we know occurred at precisely that time. In other words, the story of Atlantis is the eye-witness record of what geology has discovered about the end of the ice-age and the gargantuan tidal wave it triggered.

The Egyptians had also said there had been cities at that time, and we now know this too was true. We have not dug up the specific cities named in the Atlantis story, the ice-age versions of Athens and Sais (because no one has tried), but we have found towns built at that same time in the land between them (between Turkey & Syria).

All this happened 11,600 years ago. But there is still more to the story.

The size of animals and human beings underwent a dramatic change at that time. About 11,600 years ago, all creatures on the earth were "mega-fauna" or giant-sized animals. The giant Irish elk, for example, had antlers so big that the "wing-span" was over twenty feet. The elk had to walk side-ways through the forest in order to keep its antlers from breaking off. A similar dilemma faced the famous Saber-tooth tiger. Its canines were so large that it was impossible for this giant creature to swallow its prey unless it first dislocated its jaw. The famous wooly mammoth had great curling tusks that came arcing back to within inches of its eyes. Not only did that make it harder for a mammoth to see where it was going, but any bump or battle with a foe could put out an eye or even pierce its brain. The gigantism of all these animals was clearly counter-productive.

Therefore, none of these oddities could have resulted from "survival adaptations," in spite of what the evolutionists would like us to believe. These excessively long antlers, canines, and tusks would not have benefited any animal.

But they do have one thing in common: They result from excessive ultraviolet light exposure that triggers overgrowth of bones, teeth, horns, tusks and similar calcium-based parts of their bodies. The gigantism that these creatures manifested is additional support for the Genesis references to giants on the earth in those days.

It is also explicit proof that the sun was putting out excessive ultraviolet light in the last days before the Deluge. Moreover, it proves the sun must have been evaporating up massive amounts of water in the months and years prior to the flood.

The intense downpour of rain that brought down Atlantis and Athens could not have followed the long drought, however, unless the sun had suddenly switched its output from the ultraviolet to the infrared.

In the Queensland region of eastern Australia, there is direct evidence for this kind of solar deluge transition. At a shallow basin around what had been an ancient ice-age spring, paleontologists found a flattened circle of bones of over 10,000 megafaunas, including giant kangaroos and giant wallabies, but no humans. The strange part was how these animals died. They had been crushed as if some enormous anvil had dropped on them from out of the sky. The skulls were crushed from above. The legs were longitudinally fractured as if under a bone-splitting weight. The entire field of dry bones, looking like something out of Ezekiel 37, had been pounded into the dust, smashed down into a layer about three inches deep. Yet, there had not been a single tree nor cliff in the area. Whatever had flattened these 10,000 animals had vanished.

Or had it? The paleontologists noted that the animal bones had lain in the shallow basin in a pool of fresh water immediately after deposition. The odd part is that every other detail about the setting, including the conditions of the bones and teeth, proved that the animals had died in an extreme drought that had gone on for seven full years. The spring was the only water source for miles. That is why all the animals gathered there.

Of course, we now know what must have killed the animals. They were crushed by a Deluge... of giant hailstones, just as the book of Revelation warns will fall at the close of this age (Revelation 16:8, 10, 12, 21, *cf.* Job 38:22-23). Note the biblical sequence:

> ...the sea... became (dried up) like the blood of a dead man; and every living soul died in the sea...the rivers and springs of waters... became blood. ...the sun... was given power to scorch mankind with fire. And men were scorched with great heat... the throne of the Beast and his (global) kingdom (became) full of darkness; and... the great river Euphrates and its water were dried up...

...and there were... thunders and lightnings, and there was a great earthquake, such as had not happened since men were upon the earth, so mighty an earthquake and so great ...and the cities of the nations fell... and every island fled away, and the mountains were not found.

And there fell upon mankind a great hail out of heaven, every stone about as heavy as a cannon-ball... the plague of the hail... was exceedingly great...

...that great city which reigns over the kings of the earth...Babylon, that mighty city (he) cast it into the sea, saying, "Thus, with violence, shall that great city, Babylon, be thrown down (into the sea) and shall be found no more at all. (Revelation 16:3-4, 8-12, 14, 18-21, 17:18, 18:10, 21)

Have you entered into the store-houses of the snow, or have you seen the storehouse of the hailstones that I have reserved for the time of Tribulation, for the day of pounding-down and war? (Job 38:22-23)

It will be as it was in the days of Noah, when the heavens opened up and the rain and hailstones fell, and there was an exceedingly great earthquake and tidal-wave, and the islands were not found, and the cities of the nations' fell, and Atlantis, that great city which ruled over the nations of the world, was thrown down, cast into the sea, and has been found no more at all.

Babylon the Great seems to have a connection with Atlantis, an empire located not far from Florida, in front of the Americas. Those who may believe America is not going to be caught up in these events of the end of this age, should bear in mind that Atlantis was the global merchant empire of its day, just as America now is. And just as Atlantis was the greatest military power of its time, so now is America. And just as Atlantis was attempting to pacify the Mid-East, so also is America.

There were other Israelite accounts of the Flood. For example, note that the *Book of Jasher* includes the great tectonic earthquake and solar darkness that we now know must have accompanied the cataclysm that ended the last ice age when Atlantis sank:

> And on that day, the Lord caused the whole earth to shake, and the sun darkened (the solar spectrum descended into the infrared a week before the Deluge (see below)), and the foundations of the world raged (producing violent rumblings within the earth), and the whole earth was moved violently, and the lightning flashed, and the thunder roared, and all the fountains of the earth were broken up (lightning and thunder claps occurred as all subterranean springs and geysers exploded), such as was not known to the inhabitants (before) (that is, humanity had never witnessed any of these great upheavals before); and God did this mighty act (all these upheavals erupted at once just before the Deluge), in order to terrify the sons of men, that there might be no more evil upon the earth (the author of Jasher gives his viewpoint). And still the sons of men would not return from their evil ways. (What follows at this point is taken directly from Genesis and was inserted by the author of Jasher) And at the end of seven days, in the 600th year of the life of Noah, the waters of the flood were upon the earth. And all the fountains of the deep were broken up, and the windows of heaven were opened, and the rain was upon the earth 40 days and 40 nights. (*Book of Jasher* VI:11-14)

Curiously, the author of the *Book of Jasher* does not mention the *Nephilim* or the giants, but he does acknowledge the corrupt evil-doers in the world at the time of the flood. He thinks the reason the sun darkened, and the global earthquake occurred was to frighten mankind into halting its evil ways. Apparently, he had before him some older tradition that the sun had "darkened" its spectrum (*i.e.* toward the infrared) seven days prior to the Deluge, and on that day, there had also been an extremely violent shaking of the whole planet, which continued to rumble loudly until the rain began. He was at a loss to explain why all this happened a week before the flood itself. So, he imagined that God was sending these events in order to scare sinners into repenting.

We can now see that a solar coronal-mass-ejection, which can carry a tremendous amount of energy, may have slammed into the earth, causing a massive shaking of the entire planet. Solar spectral change is often related to such events. The author of the book could not have known that all the events he describes are scientifically credible. If we ignore his speculations about God's motives, what remains is a very straight-forward account of a severe solar spectral fluctuation and its geological effects.

He is describing this solar event from historical records he has read, as opposed to oral traditions. His version contains consistent credible details one would not expect from an orally-transmitted myth that had mutated badly in the re-telling.

More than that, the scientific data about the events of the ice-age, which we have been examining, gives us confirmation for historic veracity of these details. That is, the evidence of the stones and bones demonstrates that humanity had actually witnessed such solar and geological events over the past 25,000 years. That means that we had seen such things happen, and that (if we had the ability to write it down) we could have left an eye-witness account of them. Indeed, if we could have left such an account, we surely would have, given the awesome scale of the cataclysm.

So, the accurate technical details of the *Book of Jasher* account imply that it was based upon material from an early written version of the upheaval, most likely penned by a survivor or someone who had heard the story from an eye-witness.

That raises the question: When exactly did mankind learn to read and write?

The Egyptian priest stated rather confidently that, although the Greeks were just then beginning to re-learn "the use of letters" after forgetting much of ancient history, the Egyptians had written records in their temples that reached all the way back to the time of Atlantis, that is, to 11,600 years ago. But many Egyptologists continue to deny this. Like all the other guardians of ancient history, they insist mankind was illiterate until about 5,400 years ago, when the Sumerians supposedly invented writing.

Anyone who has studied ice-age cave art will find it hard to believe those artists were illiterate. The caves are covered with "symbols" that certainly look meaningful. Egyptian hieroglyphs used single symbols to represent entire words, and they were likewise written on walls, with sentences going in every direction. Although we cannot yet decipher them, that hardly justifies the assertion that they cannot be writing. The real problem is that scholars are too invested in the "history began at Sumer" myth.

The matter might have remained unresolved but for the chance discovery of the lost Atlantean-era stone structures and settlements of Turkey and Syria. One of these, Gobekli Tepe, is dated 11,600 years ago. (Mann 2011, pp. 34-59) Some sites date 3,500 years before that, which in theory, was well back into Atlantean times, if not before that empire had arisen. At least eighteen towns, or other structural sites from that ancient era, have been unearthed. (Mann 2011, pp. 42-43) The dates range from 6000 B.C. (8,000 years ago) back as far as 13,000+ B.C. (over 15,000 years ago). That makes some of these sites nearly three times as old as Sumer.

Gobekli Tepe reveals that humanity had literary capability 11,600 years ago. The site includes some cleverly carved stones. Hard granite has been chiseled away to reveal bas relief images of various precisely-rendered animals and symbols. This method of detailed, up-raised carving is very difficult and quite sophisticated. Accordingly, it implies a long prior tradition, not only for the development of such stone carving, but for the development of the symbols themselves. One can easily infer many centuries for the invention of the carving, symbolic, and stone-construction capabilities at Gobekli Tepe.

That would date their origins long before Plato's 9,600 B.C. date for the sinking of Atlantis, a date, remember, that is supported by the geophysical upheavals we now know occurred at that same time and place.

Moreover, the Gobekli Tepe symbols look surprisingly modern. For example, on just one stone pillar we can see a string of symbols (looking just like a word) that appears roughly like this: "...C HC HΔ..." On the same line, but around the corner on the pillar are a few I and U and H shapes and one I symbol. As you can see, all of the symbols look like letters in our alphabet, including the Δ which is the original shape of our D. There are no symbols in the group that do not relate to our alphabet, or its ancient origins in Phoenician writing (which is similar to Hebrew of the same time).

What are the odds that all these different symbols would be "alphabetical" and strung together as if in words? Keep in mind that there are no other symbols but these in this line of symbols.

Now you should know that as we go back in time some of these symbols had a different usage than we now give them. For example, the H used to be a form of the E. The C might have been more like an S, and a symbol like the I shape once stood for our Z sound. The I evolved into our I and J and maybe W sounds. And the U became U, W, and V. And the V (which is only partly shown but implied by the visible right half on the pillar) became a W, F, B, or it could even have been an N in Greek.

You may think all these variations seem excessive. However, all of these changes in alphabets are fairly well-established by archaeologists.

Let's pretend we could read what is on this Gobekli Tepe pillar. What might it say if rendered the way those symbols were later understood? One possibility could be something about like this: "W'H'SED ZEEVIV'ZZ." You might find that a bit obscure, but a student of Semitic language might read it as "W 'Hesed Zebub 'zz," meaning "And even (you) bow down (to) Zebub 'zz! (= Beelzebub? See *Strong's #1176* (לעב בובז))."

It is not clear whether this reading would be a command to bow to Beelzebub or a criticism of people who do. Another reading might be, "Even (you) will be brought down to (be consumed by) flies!" This latter interpretation is like the Genesis statement that, "Dust thou art and unto dust (flies) thou shalt return!"

For our purposes here, it only matters that it is possible to read the line of symbols as a relevant Semitic sentence, especially because archaeologists are convinced this pillar was at the heart of some sort of religious structure. (Mann 2011) In other words, we have made a reasonable case for Semitic religious writing at an arguably Semitic religious site (near Edessa) built right after the reputed sinking of Atlantis, about 11,600 years ago. This is sufficient to show that someone from that time might have been able to leave a written record of the events in a language that later Semitic peoples might have understood.

If we identify the Deluge of Noah with the flood that sank Atlantis, then the story of Noah also adds support to the idea of a written record. As theologians and scholars have long realized, Noah's account reads exactly like a captain's log of a voyage. Thus, it strongly suggests Noah had been writing a log on board the ark. It too could have been in a Semitic language, given that Shem, for whom Semitic peoples are named, was said to be with Noah on the ark, when there was only one language. (Genesis 9:18, 11:1-10)

Virtually every ancient culture told Deluge stories. The Chinese, for example, even called their Noah character "Nuwah" and gave a similar account of his ark. Even if we did not have the Genesis version, we could readily reconstruct much of the Genesis story from all the other surviving versions. And, as we have noted, the global sea-level suddenly rose 375 feet, virtually overnight,

11,600 years ago when Atlantis sank. Many giant species (woolly mammoth, mastodon, woolly rhinoceros, giant dire wolf, etc.) were wiped out at that time in a world-wide mass-extinction event. So, the Genesis Deluge account is (or ought to be) a credible record. Were it not for the religious aspect of the story, it might well be accepted history today. (II Peter 3:3-6)

All of this reinforces the credibility of the ominous account in the *Book of Jasher* about the sun darkening just as the earth was being struck by a global super-quake. When we remember that the solar wind arrives days after a change in the solar spectrum, the concurrence of the two events on the same day tells us that the sun must have had a massive coronal ejection a few days before the color turned red and the light dimmed. With our modern satellite sensing capability, we ought to be able to get some advance warning of the approaching CME. But even if we did not, the *Book of Jasher* indicates there would still be a seven-day warning of the deluge of hailstones that is predicted to end this age.

That pounding with hail is the "treading of the winepress" of Armageddon, when Christ returns in "the clouds of heaven," following the time "the sun will be darkened." But these things come after "the tribulation of those days... shortened for the sake of the elect" (Revelation 14:14-20, 16:12-16, Matthew 24:22-30). What sort of "tribulation" can this be that a solar deluge seems a relief? The plot must thicken for us to answer that question...

For the Deluge of Noah was not the only time God had to intervene on a global scale to wipe out races of giants. In the next chapter, we will investigate a story from the ancient *Book of Jubilees* about something that devastated the whole of North America.

Chapter Bibliography

Bartholet, Jeffrey. 2012. "Swept from Africa to the Amazon." *Scientific American*, February: pp. 44-49.

Brown, Michael H. 1990. *The Search for Eve*. New York: Harper and Row.

Felix, Robert W. 19097. *Not by Fire but by Ice: Discover What Killed the Dinosaurs...and Why It Could Soon Kill Us*. Bellevue, Washington: Sugarhouse Publications.

Gribbin, John, and Mary Gribbin. 1996. *Fire on Earth: Doomsday, Dinosaurs, and Humankind*. New York: St. Martin's Press.

Mann, Charles C. 2011. "The Birth of Religion." *National Geographic Magazine*, June: pp. 34-59.

Marean, Curtis W. 2010. "When the Sea Saved Humanity." *Scientific American*, August: p. 55.

Morris, J. D., and J. J. S. Johnson. n.d. "The Draining Floodwaters: Geologic Evidence Reflects the Genesis Text." *Arts and Facts*, pp. 12-13.

Plato. 2015. "Timaeus." *In The Dialogues of Plato: The Republic, by Plato*, translated by Benjamin Jowett. Seattle, Washington: Andesite Press, an Amazon Company.

Quayle, Stephen. 2002. *Genesis 6 Giants*. Bozeman, Montana: End Time Thunder Publishers.

Raeburn, Paul. 2012. "The Moon is Full of Surprises." *Discover Magazine*, March: p. 26-28.

U.S. News and World Report. 2009. "Adam, Eve, and the Genome." Mysteries of Science Edition: pp. 45, 47.

U.S. News and World Report. 2009. "Retracing Our Steps Back to the First Humans." Mysteries of Science: pp. 48-49.

U.S. News and World Report. 2010. "The First Sacred Place." October: pp. 16-19.

Velikovsky, Immanuel. 1950, 1967. *Worlds in Collision*. New York: Dell Books.

Chapter 3

The Sword of the Lord

The book of Revelation says a deluge of ice finishes the seventh and last plague, when the cities of the world will fall, and the mountains will not be found. (Revelation 16:17-21) That global pounding down with hail may be what Revelation elsewhere calls the "treading of the winepress" of Armageddon, when Christ returns in "the clouds of heaven," following the time when, "the sun will be darkened and the moon not give its light." But this comes after "the tribulation of those days... shortened for the sake of the elect... or there would be no flesh saved alive."[1]

What "tribulation" is so deadly that it must be cut short by a global deluge of ice?

For the answer, we must look to the past. The Deluge of Noah was not the only time God had to intervene on a global scale. We need to investigate an ancient story, but barely mentioned in the Bible, that involved God rescuing mankind before the flood. The Bible sums up most of what it says about that pre-flood world in chapter six of Genesis:

[1]Isaiah 13:10, Joel 3:15, Matthew 24:22, Mark 13:20, Revelation 14:14-20

...when the (line of) Adam ("the bloody faced ones") began to multiply on face of the earth, and daughters were born to them, sons of God saw the daughters of the (line of) Adam that they (were) fair; and they took them wives of all they chose. And the Lord said, "My Spirit will not always strive within Adam because of their sinning; (for) he is flesh and his days (after this painful corruption of his line) shall be 120 years.

The giants (*Nephilim*) were on the earth in those days, and also after, when sons of God came to the daughters of the (line of) Adam and they bore (giants) to them. These (giants) were "mighty men" (or *Gabarim* of Asar or Warriors of Osiris) from ancient-time, men of renown (or "from the ancient time of those sons of Enosh of the line of Shem"). And God saw the wickedness of the (line of) Adam (had become) great on the earth, and that every imagination of the thoughts of his heart (had become) only evil all the day. (Genesis 6:1-5)

If these "sons of God" were normal human males, they would not have produced odd offspring. The whole point of the passage is that these unions resulted in giants (as parallel accounts in other cultures agree). Since the mothers were normal, the term "sons of God" (*ben a h'Elohim*) must refer to beings who were not normal human males, but somehow genetically different. When "sons of God" is used in the Hebrew Scriptures, it usually refers to beings apparently created by God, but different genetically.

So, these giants are the product of fallen beings that "took" wives that they "chose" (whomever they wanted). This needs explanation. In the world of the Bible, a daughter was given in marriage, not taken. This passage is indicating that the women in question were being abducted ("taken") and then seduced or raped. We get further insight about this from the parallel passage in the pseudepigraphal *Book of Jasher*. It is a reflection of what some ancient readers thought Genesis meant:

> And their judges and rulers went to the daughters of men and took their wives by force from their husbands according to their choice... and (they also) taught the mixing of animals of one species with the other... (Jasher 4:18)

Here it is clear that the text means they "abducted and raped" these women, but it adds that these women were already married (making the crime worse). We are also told that these "sons of God" were "judging and ruling" humans. So, these are not just human beings, but fallen beings that have taken over rulership of at least part of the planet. Evidence for the knowledge of advanced genetic manipulation technology is also given in this text. This is what Genesis calls the "corruption" of the animals.

Genesis 6 emphasizes the name "Adam" (literally, "bloody-face"). It is speaking of two time-frames. There is the initial time, when daughters were first born to Adam and his sons. We are given the cryptic reference to "120 years," which in retrospect, from our self-centered modern experience, we simply assume refers to our own modern life expectancy. But, this text is not about us. It is about Adam, as it plainly says: "My Spirit within Adam (which God breathed into Adam's nostrils) will not always strive in their sinning" (will not forever endure his daughters' straying with fallen beings). The text specifies "in Adam (the first man himself); he is flesh and his days (after these abductions began to happen) will be 120 years." In other words, this corruption of Adam's family occurred 120 years before Adam died, *i.e.*, around the mid-point of the pre-flood age.

This explicit reference in the text to Adam himself is no accident. We are then told in the next verse that another corruption of daughters of Adam's line came "also after" when "sons of God" again seduced women and impregnated them. It was a return of the sexual corruption of humanity that first began in Adam's time.

We need to make an observation about the translation. This Genesis text never says "the" sons of God. This implies that it was only some, not all these beings were seducing or raping these women. Not all these beings were fallen. The insertion of "the" by translators has created the false impression that all these beings of the time were abducting women. The Hebrew text does not say that.

Obviously, we need to realize that these "sons of God" are not the kind of purely "non-material" beings many have assumed them to be. At the very least, they seem to have the ability to transform themselves into a material form or in some way fully manifest in this three-dimensional world. In the New Testament, we are told that it is common for people to "entertain angels unawares" and interact with angels as if they were ordinary human beings (Hebrews 13:2). That clearly implies that they would be able to interact physically with them.

The fallen sons of God corrupted not only women, but all living creatures through genetic manipulation. All the creatures of God were good (I Timothy 4:4). Attempts to change them corrupted them, causing short life expectancy and violence. The fallen beings knew this. They could not improve on God's creation. Their goals, Jesus said, are "only to steal, to kill, and to destroy" (John 10:10). Fallen beings were trying to corrupt God's creation. We know how God reacted to this situation. Note the "wipe the blood from the face" pun:

> And the Lord regretted that he had made the Adam on the earth, and he was grieved at his heart (the Lord or the Adam?). And the Lord said, "I will wipe away the Adam ("bloody face") whom (or "the man Asar (Osiris)") I have created from the face of the earth, from Adam to the beast and to the creeping thing and the fowls of the air, for I regret that I have made them." But Noah found grace in the eyes of the Lord...

> The earth also was corrupt ("marred") before God, and the earth was filled violence. And God looked

upon the earth, and, behold, it was corrupt; for all flesh had corrupted his way upon the earth. And God said to Noah, "The end of all flesh has come before Me, because the earth is filled with violence through them. And, behold, I will "corrupt" them (the corrupters) with (water which will "mar" the surface of) the earth... And of every living thing ... two of every sort shall come unto thee (in the ark) to keep them alive. (Genesis 6:6-13, 7:19-20)

These corrupt beings endangered all flesh. God was determined to remove them but preserve the uncorrupted. Would God wait until the flood? We are told God "changes not." Could God have already acted before the Deluge to remove giants and protect life from corruption? Genesis cites two corruptions of women by *Nephilim*. According to the apocryphal *Book of Jubilees*, there was a prior destruction of the *Nephilim*:

...and he called his name Jared, for in his day the angels had descended upon the earth, those that are called Watchmen... The angels of the Lord saw them (the women) ...that they were beautiful... and took... wives... and they bore them sons, and these were giants... All flesh corrupted its way, from men to animals... and began to devour each other (had become cannibals) ...And He was greatly enraged about the angels He had sent down upon the earth, saying He would root them out of all their powers (strongholds?) and that we (good angels) should bind them in the depths of the earth... and against their children, the *Nephilim*, there came a word from the face of the Lord, saying that they (the *Nephilim*) should be slain with a Sword and be (thereby) removed from under heaven... And He sent into their midst His Sword... and they all fell upon the Sword and were destroyed from the earth ...And

He destroyed ALL their places, and there was not left a single one of them... (Ethiopic Book of Jubilees 4:15, 5:1-2, 5-6, 8, 10)

All were killed by this "Sword of the Lord." What "Sword" could this be? Did some kind of Heavenly Sword slay all the *Nephilim* and "destroy all their places" before a later destruction of other giants by the waters the Flood? This could explain why the *Nephilim* seem to have disappeared for a time after being on the earth during Adam's life, and then were on the earth again afterwards, when they returned and corrupted the daughters of Adam's line a second time. (Genesis 6:4)

This would mean that God cleansed the earth of the *Nephilim* giants twice. The first time God had used something called "the Sword of the Lord" and the second time God used the Solar Deluge of Noah. The first cleansing had indeed killed all the *Nephilim* on the earth, *Jubilees* says, but then we find there were still some fallen beings left who apparently escaped. They returned, and this time, having learned how to corrupt more efficiently, the only way to cleanse the earth of them (and the genetic corruption they had caused) was to wipe away everything with a global deluge.

What, then, was the Sword of the Lord?

Was it a comet? Many have proposed that comets were seen during upheavals on the earth, were blamed for these disasters, and identified with whatever god or goddess ancient cultures at the time were worshipping. In other words, they claim it was a coincidence that a comet was passing by at the time some disaster chanced to be happening.

I can attest to how impressive a comet can be. At midnight on March 23-24, 1996, I stepped outside to see Comet Hyakutake passing overhead near its close approach to the earth. I have used a small telescope to watch the skies for over 50 years. I didn't need it that night. Comet Hyakutake (see picture at beginning of next page) was so large I could stretch out my arms and scarcely span the whole extent of the comet as it drifted silently overhead. It was huge. I could not see all of it without moving my head from left to right across virtually the whole visible night sky. Nothing I have ever seen in the natural heavens in half a century has ever come close to what I witnessed that night. It was awesome. Comet Hale-Bopp, perhaps the most massive passing  comet studied by modern astronomers, was so far away at its closest, it was a disappointing tiny little blotch by comparison to magnificent Comet Hyakutake.

The comets we see are, at their largest, 25-50 miles in diameter. But the mass is not as great as that size at first suggests. Much of a comet is water and frozen gas or a fine dust that is easily stripped away by the solar wind. For a comet to kill anyone on the earth, it would have to take a suicide dive into our atmosphere, destroying itself in a fireball whose sky-blast would scorch everything directly below.

There is a theory that on June 30, 1908, a small comet did just that, when it exploded in the sky over Siberia, killing a few caribou and leveling a forest like the detonation of an atomic bomb. (Hogenboom 2016) However, had it impacted a few hours later, it could have scored a bull's-eye hit on London, which was then the financial capital of the world. That would surely have re-written the history of World War I and everything that followed. This alternate scenario shows what a comet impact might do to a civilization.

These secondary effects on history are hard to assess. So, it is hard to match that kind of localized comet impact with the mythic legend of the "Sword of the Lord," something that arguably happened between about 18,000 years ago and 12,000 years ago, but which continued to leave its imprint in literature composed around 2,000 years ago.

The *Nephilim* were not concentrated into an area as small as metro London. Yet, the *Book of Jubilees* said "all of their places" were destroyed by the Sword. Several locations on the earth were wiped out completely. That was a massive destruction. Mankind survived, but none of the *Nephilim* of that time survived. The fallen beings had to regenerate the race of giants afterward (Genesis 6:4).

What could have caused so thorough a destruction of these giant beings?

We know normal-sized humans could hide in caves, whereas we have found that the giants were often too large. Also, their bulk required them to eat an enormous diet each day. By contrast, humans can survive on relatively little but bread and water for weeks on end. Giants would have had trouble maintaining their body temperature during climate change. Humans are far more adaptable.

We may conclude, then, that we humans have several survival advantages over giants in a global cataclysm. But what kind of cataclysm was it?

Perhaps we could learn more if we could isolate the time of the event and find some scientific clues that would help define what happened. We can roughly identify the time of this event as happening somewhere during the last half of the last ice age, which ended with the solar deluge 11,600 years ago. That places it c. 15,000 ±3,000 years ago, or roughly around 13,000 B.C. It was an extinction-level event, but not primarily a flood. It seems to have involved some significant cosmic body or celestial phenomenon that could be described as "the Sword of the Lord."

It happens that such an event is actually now known to have occurred about that time, and it has been investigated sufficiently that we can map out its details fairly well.

First, you may recall that during that time there was a huge slab of ice nearly three miles high sitting astride the Midwest of North America. Centered in the upper Great Plains, its frozen water had been extracted by the sun from the oceans of the world.

Also, any seasonal melt-off from that glacier normally went in one of three directions: either it went down the Missouri River into the Mississippi and on into the Gulf of Mexico, or it drained into Hudson's Bay, or it had to fill the basins of the Great Lakes and run out the St. Laurence River to the Atlantic.

If any comet had exploded over North America, it would have melted the upper layers of that glacier and should have generated flooding in these three areas. But it would also have overflowed the river banks and lake beds of its drainage basin, which would have left behind a significant flood layer in the sediment. That flood layer should be filled with black ash from the flash incineration of vegetation caused by an exploding comet. Moreover, it ought to have a layer of diamond dust, which constitutes about 10% of the tail material of the typical solar system comet. Finally, this layer should cover an area large enough to have killed every living creature over a portion of the earth.

And it does... Studies have found all of the above characteristics at the right time, about 15,000 years ago. Almost anywhere in North America, in fact, one can dig down about eighteen inches and find this black burn-layer. Traces of it have been detected over such a vast area that it is now estimated to have enveloped a third of the planet.

The scientific reconstruction of this colossal calamity seems to point to a huge comet exploding above what is now Chicago (but this may just be the drainage pattern). It was a sky-burst which instantly incinerated every exposed bit of combustible organic matter beneath it and even at some distance. The ice directly below could have entirely melted, and a lot of steam would have scalded off the upper portions of the rest of the glacier. This super-heated vapor would have been blasted away by a viciously hot wave flowing at hypersonic velocity out from the explosive center of the event.

Of course, some of this new atmospheric moisture could eventually be blown back over the glacier and partly rebuild it. But the blistering tidal wave of white-hot steam and dust would have sand-blasted the land of all animal life not protected by caves or fortuitous canyons, and then only at a great distance from ground zero.

Had there been Atlantean colonies in America yet (it was nearly 4,000 years before the kingdom of Atlantis was reputed to have sunk), they too would have burned. And we can be sure, if North America had been the land of the *Nephilim*, they would have all perished in a few minutes.

This was obviously a far more impressive comet than the one that hit in 1908. Indeed. It would have to have been larger than any comet ever measured by modern science. If it had been seen in the night sky before the impact, it would have dwarfed even Hyakutake. Its head would have glowered above, larger than the full moon, and the end of its tail would have been beyond the far horizon. It would have been difficult to find a part of the sky it did not cover. Its enormity must have forever haunted the nightmares of all who survived, for, seconds after it passed over their distant mountains, it disintegrated with the light of million suns. The whole earth shuddered. A deep roar ascended from the ground and a cold gale of wind was sucked toward the impact, then halted and reversed, returning as a flying wall of steam and boiling mud.

Had giants walked in the shadows of the great glacier, they died that day. Those not vaporized were melted. Those not melted, were burned. Those not burned, were swept off their mighty legs and carried aloft like rag dolls until they shattered against mountains or found themselves hurled into the far ocean, a feast for sharks.

When survivors of Jared's day investigated, they sailed into empty harbors and surveyed landscapes barren of life. All the abodes of the giants lay in ruins. Little remained but the lingering legends told by those they had once terrorized.

All the ice age archaeological sites of North America, all its remnants of human occupation, date from after that time. (Pringle, *The First Americans* 2011) In spite of decades of diligent exploration, with many generations of American archaeology students sifting through local digs, only a handful of ice-age human sites have been found in the Americas. (Pringle, *The First Americans* 2011)

Skeletons of giants are not considered "accepted" finds. Several of those have been unearthed in the Americas, but they are not included as "scientific" discoveries, apparently because giant humanoids might lend credence to the Bible. (Quayle 2002)

We will limit ourselves here to the generally accepted, but nevertheless revealing, finds. By the way, archaeologists call the post-cataclysm survivors from 15,000 years ago to the time of the end of the ice age in the Mideast, the Natufian culture. Their sites range from Syria to Israel, the very center of Jared's pre-flood kingdom. (Mann 2011, pp. 42-43, 49, 56)

All dates are approximate and based upon techniques (typically the Carbon-14 levels of organic material found in close proximity with stone tools) that have a range of error up to *c.* 1,000 years. However, because multiple dating samples are taken at each site, the average (or the median) date is usually the one used (with any extreme measurements normally excluded), and it would have less of an error range than any uncorroborated date. Here is a survey of all sites in the Americas, with the recent finds first, going back to the earliest:

Off the California coast, on Santa Rosa Island, which was nearer the mainland during the ice age, small stone arrowheads about the size of a large postage stamp, were found in a context dated to about 11,800 years ago. (Pringle, *The First Americans* 2011, pp. 39-40, 44-45) These points resemble others that have been found in eastern Asia and that have been dated to around 15,000 years ago. Putting the two finds together, one could speculate (the term scientists prefer is "the similarity suggests") that hunters first migrated to the Americas from Siberia *c.* 15,000 years ago. (Pringle, *The First Americans* 2011, pp. 39-40, 44-45)

Beneath the 90-foot deep Alaskan muck lying semi-frozen across the far north, through which everyone agrees any early migration had to pass, a single stone spear-point was found. This spear-head must date before the frozen muck was deposited, when continent-sized tidal-waves of mud and volcanic ash over-swept Alaska and Siberia, in a matter of minutes. Giant mammoths were found torn in half and flash-frozen in the muck. At this same time in the Arctic Sea north of Russia, the New Siberian Islands were instantly formed by a horrific tidal wave of mammoth, mud, ash and twisted tree trunks.

Volcanic carbon makes it hard to date such events. The sinking of Atlantis and the ice melt of North America, which we have identified with the solar deluge of Noah's day, took place 11,600 years ago. But the volcanism and violence of the Alaskan-Siberian events are more likely to result from a sudden cosmic impact, rather than a solar deluge.

Near Clovis, New Mexico, archaeologists in the 1930s began finding additional spear-heads shaped something like the lone one found underneath the Alaskan muck. These could be dated in their context to 13,000 years ago, pushing the human settlement of North America back 1,400 years. (Pringle, *The First Americans* 2011, p. 40) It is thought the spear-heads were fashioned to bring down large mammoth. For decades, the "Clovis People" theory of the settling of the Americas was the dominant view of anthropologists.

Ancient arrow-heads like those at Santa Rosa but dating to 14,000 years ago have been dug up in Oregon. (Pringle, *The First Americans* 2011, p. 40) Mesa Verde, Chile, in South America is the next site to push back the age of man in the Americas, dating 14,600 years ago. It is the oldest of a number of sites that have been found near the Pacific Ocean coastline. Older sites may lie submerged offshore. (Pringle, *The First Americans* 2011, pp. 38, 41)

Finally, there is now a treasure-trove of 19,000 chiseled points and tools being unearthed at a site near Buttermilk Creek near Salado, Texas. (Pringle, *The First Americans* 2011, pp. 36, 38, 41) (Waters, *et al.* 2011) (Pringle, *Texas Site Confirms Pre-Clovis Settlement of the Americas* 2011) With meticulous multiple dating to support it, the site may have been occupied as early as c. 15,500 years ago. Nothing earlier is known. (Pringle, *The First Americans* 2011, p. 38)

New optically-stimulated-luminescence (OSL) dating of sand layers, from the great North American central plain along the easterly flanks of the Canadian Rockies, indicates an ice-free corridor suddenly opened *c.* 15,000 years ago. (Pringle, *The First Americans* 2011, pp. 40, 44-45) This dating method can date the sand directly, without resorting to the vagaries of C-14. The dates ranged from 14,000 to 15,000 years ago, but it would have taken an estimated 1,000 years to create these dunes after the ice melted. (Pringle, *The First Americans* 2011, p. 44)

So, it was the melting of the ice corridor down the middle of the glacier that opened the door for East Asians to migrate through

Alaska and down into Canada and the American plains to Texas, New Mexico, and eventually South America. This sudden glacial melt-off would have occurred *c.* 15,000 years ago. (Pringle, *The First Americans* 2011) This seems to show that there was no resistance at all to the new immigrants, who moved south almost as fast as they could walk, not stopping until they reached Buttermilk Creek in Texas.

It is unlikely that the hunters would have entered this strange new land if the big game they were stalking had not been fleeing before them into the Americas and were moving south just as fast as they were. Hunters follow the migrations of their food supply. Mammoth, like any other big game, hold a territory that they defend against competitors. The migration of mammoth into America from Asia indicates that the native population was not defending itself. Man, and mammoth entered the new land equally fast and seemingly met no resistance at all.

Why were there no defenders of this supposedly lush landscape? Because they had been burned out. But why did the immigrants not stop until they reached Texas?

The newcomers, both human and animal, were ultimately dependent upon a base of trees, grasses, and other vegetation. But if these had been incinerated, the landscape they encountered was not lush, but comparatively barren and desolate. There were no trees in the corridor with leafy branches for the mammoth to munch.

So, the herds of mammoth fled before their Asian pursuers, eating what little scrub brush had taken root in the ruined land. Hunger drove man and beast ruthlessly southward until they reached a place that had weathered the cataclysm well-enough to support year-round occupation: Places like Texas, New Mexico, and California.

It was all about survival. Had they been able to make a life in the northern corridor or plains, they would have set up a year-round camp there. But as my Scandinavian ancestors discovered in the 1880s and 1890s, the Great Plains can get brutally cold in the winter, and there isn't much there but sod until you plant a crop.

These new invaders were hunters like Nimrod, not farmers like Cain, nor herders like Shem. They kept warm with slain mammoth skins on their back. They cooked meat over fires fed by mammoth blubber. The courage to stalk and kill an animal that could step on them like a bug was how they stayed alive. It was only natural that people with this kind of fearless determination would dare to be first to probe the land that God had just burned over.

But where did that lengthy open corridor come from?

If one looks at the before and after view of North America, the southern melted off areas are generally under the "exploding comet" that had hit *c.* 15,000 years ago. But why would the northern part of the corridor have melted at the same time?

There is a principle in science called *Occam's Razor*. It says that when faced with a complex set of possibilities, the simplest answer is most likely to be correct.

In this case, is it more likely that two or more independent events created the corridor in the heart of the great glacier, or that one event did it? The answer, of course, is that one single event is the more likely cause.

Likewise, if a comet event struck in North America and opened up the corridor through the very heart of the great glacier, did a completely separate impact strike deep in Siberia and sweep its mammoth northward into the Arctic Sea to form the New Siberian Islands, driving the survivors eastward? Or, were all these things related?

Yet, knowing some kind of large-scale fiery cataclysm was involved, we should ask if perhaps the cause was not a single comet exploding in one big blast, but some long and narrow destructive mechanism that fell upon the ice in a single calamity.

We need not invent such a mechanism. On July 16, 1994, 21 fragments of the Shoemaker-Levy Comet left a long string of impacts on the planet Jupiter. The break-up of a comet into a string of impacts may also have occurred 65-million years ago when the dinosaurs died. (Gribbin and Gribbin 1996, pp. 25-41)

The dinosaur killers are especially revealing. They fell in a string that stretched from the Yucatan of Mexico, through a crater east of Manson, Iowa, to one in Alaska and finally, several craters in Siberia. (Gribbin and Gribbin 1996, pp. 25-41) Check it out on a map. It is virtually identical with the track where the ice corridor suddenly opened up about 15,000 years ago.

In fact, if one were a believer in a recent creation a few thousand years ago, it would be tempting to argue that the aged string of impact craters that define the dinosaur extinction were the very ones that actually did open up the ice corridor. The parallel paths are that close.

On the other hand, the similarity may give us an insight into just what happened that fateful day when North America burned, for it turns out that by mapping the path of the craters, scientists have determined the time of day and the time of the year when the dinosaurs perished. (Gribbin and Gribbin 1996, pp. 25-41) Because the 15,000-year ago comet impacts followed almost the exact same path, their timing data ought to be nearly identical.

According to the studies of the dinosaur crater distribution, the comet broke up over South America, entering earth's atmosphere around mid-day. British astronomer John Gribbin suggested the comet fragments that wiped out the dinosaurs (and 70% of all life on earth) had descended out of a "clear blue sky" in early June. (Gribbin and Gribbin 1996, p. 33)

This might not be exactly correct. The earth, it was found, was tilted toward the sun in the Northern Hemisphere at the time of these impacts. (Gribbin and Gribbin 1996, p. 33) Technically, that would be a few days either side of the Summer Solstice on June 22. It is just as likely that the impacts occurred in early July as in early June. The closer to the Solstice itself, the better the impact craters fit the calendar requirements.

This hints at another way to identify the specifics of the 15,000-year ago event. Afterward, the date would have been kept by survivors as a great holiday. The cannibal *Nephilim* that had threatened to devour mankind had been (temporarily) annihilated from the earth. The abduction and rape of young women had ceased. The domination of the evil angelic overlords had ended. And the humans who had huddled in fear on the far side of the earth that terrifying night had not died, but had been saved from their horrible, insatiable enemies.

They had come through the Fires of God, as if baptized by fire, and they and their world were reborn that night.

How could the survivors not have forever celebrated such a miraculous rescue?

And so, they did. There is an ancient holiday celebration around, but not quite on, the Summer Solstice. Known as "St. John's Day," it is kept two days later, on June 24th. On that day, ancient peoples had set bonfires and had taken turns jumping through the flames. (Gaster 1966, p. 36) This may seem risky, but it makes sense if they were celebrating "divine" fires that killed the *Nephilim*.

The "St. John" is John the Baptist. He received this honor for two reasons. First, he symbolized baptism by fire and water. Second, his birth was assumed to be six months before Jesus was born (Luke 1:36), and it was wrongly believed that Jesus was born at midnight on the 24th of December. So, this celebration in June was at midnight in the Eastern Hemisphere, at the very time of the final *Nephilim* high noon in the Americas.

Of course, it all makes perfect sense. The tradition in some lands even has the people building up a wicker or wooden "giant" to burn at midnight, when the giants were indeed burned on the far side of the world, at what was their noonday.

With all these ancient traditions in hand, we can now date the event to around noon U.S. Central Time, on June 24, about 15,000 years ago. We may even be able to propose a possible specific year, but that will take several more reports to attempt that.

The "Sword of the Lord," then, was a string of comet fragments that stretched across the sky in a glittering parade of Doom in mid-June c. 15,000 years ago. It might have been visible for a few days before it struck, appearing most vividly in the night sky of peoples in the Middle East. They saw it and wondered, no doubt, what it meant, trying to interpret this cosmic "Sword" as a mysterious sign from God.

When the Sword brought a huge flash of light and a scalding blast of wind and muddy rain from the far side of the world, they surely must have hoped it was a divine punishment of the *Nephilim*. And soon they learned that the places of the *Nephilim*, as Jubilees recalled, had been all destroyed.

Now we have an explanation for the omen-reading sages of the past, men who would watch the skies for signs from the gods and foretell doom and gloom whenever they spied a comet coming (cf. Luke 21:11, 25-28). In Israel, however, the prophets would declare that "His arm is stretched out still" (Isaiah 9-10) and would forecast the destruction of Israel's enemies just as often as they would warn of the Lord's wrath upon Israel itself.

So, the Sword of the Lord sliced through the heart of the *Nephilim* kingdom in the Americas and cut a swath through the gigantic glacier that towered over it. Not only did it cleanse the planet of the giants of that time, but it also simultaneously opened the gateway for the ancestors of "native" Americans to people the West (cf. Exodus 14:13-21).

But wait... Why should a people then living securely in Siberia, as their similar spear-heads show, decide to head north and east into this scorched, lifeless ruined land?

When people whose survival depends upon hunting suddenly migrate thousands of miles, it means two things have happened: The game is gone from where they had been living, and they had reason to believe it had moved to the new land.

The migrating hunters would have stayed where they were if they could have done so. Hunting is a labor-intensive activity and depends upon chance events. It has enough work and uncertainty in it that no one would leave a secure hunting ground to risk starvation in an untested new one, that is, unless forced to flee.

So, we can be confident that their old hunting areas had become insecure and that the mammoth had fled. The hunters had no choice but to move into the depopulated and scorched American wilderness. That must imply that Siberia was in an even worse state than the devastated Americas. What was true for the hunters must also have been true for the mammoth they followed. The mammoths' food supply--the trees whose leaves they ate--had suddenly disappeared.

Of course, we can now reconcile these details. The diagonal string of impacts that had melted the glacial corridor, just like the ones that killed the dinosaurs, did not end in Canada. They continued into Siberia, which seems to attract comet hits in June on a rather frequent basis. That is not a coincidence, as we will discover later.

So, there had been a comet strike over Siberia that had generated that devastating tidal wave of mud and ash that swept mammoth into the Arctic Circle, forming the New Siberian Islands. The surviving mammoth that were grazing in the Bering Sea region (then above water) fled in terror away from the Siberian impact, not realizing there were fleeing into a land that had also been destroyed. The hunters, having nothing but fire and mud behind them, had no alternative but to move forward after the fleeing mammoth.

No wonder people attributed these events to supernatural causes. The same events that killed the cannibal giants also miraculously opened the corridor in the middle of the great glacier and drove the mammoth and the hunters of Siberia into the Americas. (Gaster 1966, p. 36)

Native American peoples believed in a "great white father" who had led them into this new land and had provided for them. They also generally believed that they were to be "caretakers" of the land and preserve it for another people, a mighty civilization that would one day return to reclaim their lost land.

Why would they come to think this? Perhaps the remnants of giant earth-works (which the natives to this day say were built by an ancient race of giants) had convinced them that these powerful giants who had mysteriously vanished from the land might one-day return to it. This will discomfit those who believe God had told the natives to preserve the land for tall, white European settlers.

The common misconception that the mammoth of Siberia and Alaska died in autumn is based upon assumptions about when the leaves in their stomachs had been on the trees. It seems there were signs of color change on some of the undigested foliage in the flash-frozen animals.

Living in Maine, I know a tree begins to turn its leaves at a different time in the same season depending on where it is growing. Those up north turn color weeks before those in the southern part of the state. And, here in the southern part, I have personally seen the leaves turn as early as August. That would mean they can begin turning in July in the north of Maine, and that they can turn in June just south of the Arctic Circle where these mammoth had been. Moreover, the mammoth died in the midst of an ice age, not a warm period like ours. The leaves could have generally begun to turn even earlier under ice age conditions. That can reconcile a June 24th timing with the Siberian-Alaskan event.

However, the solstice was not near June 24th back then because precession of the equinoxes had moved the date of summer. Two days after the Summer Solstice would have been January 27th, a mere 15,500 years ago, if our calendar were not tied to the seasons. The Romans tried, often ineptly until the time of Julius Caesar to keep their sacred calendar's feasts tied to the "correct" season. So, the dates they honored became very confused.

The Romans had switched the months of January and February in an attempt to keep a February feast in sync with the seasons. (Richard 1999, p. 207) So, February had been the first month, and the odd part is, the original "date" of the destruction of the giants would then have fallen *c.* February 27, or the day before their "eve" of January 1st (that is, the day before our New Year's Eve).

We have always assumed that the eve of January 1st was such a blockbuster annual celebration because it was New Year's Eve. Certainly, the pagans had not made a big fuss over it because of the mistaken belief that Jesus had been circumcised on that day. Pagans counted it the 12th night after Saturnalia began. Catholics moved twelfth night to January 5-6th. After the Caesar's calendar change, the first of January represented the new-born baby of the year, whereas the eve was the dying old man of the past.

The Roman god Janus, for whom this month was anciently named, was a deity who was honored as the ancient Etruscan high god. (Turner and Coulter 2001, p. 249) He was the "god of beginning and end," and of gateways. He was shown having two faces turned in opposite directions. Sometimes one face is old, and one is young. His consort was the goddess of Secrets. The cataclysm had been the day before their holiday. (Turner and Coulter 2001, p. 249)

But, this was meant to honor the second month of their calendar, not the first. We assume they were keeping New Year's. What if it was not about the calendar, but honored the end of *Nephilim* rule and the "rebirth" after earth's scorching by the comet that broke up 15,000 years ago?

They celebrated the day that followed the catastrophe, that is, on the last day of February. Yet, before all these calendar changes, when February was in Spring, they had late October in our January 27th position, at *Nephilim* Summer Solstice time. That places it near their Halloween, the night evil dead *Nephilim* stalked the land seeking revenge.

The cataclysmic break-up of a huge comet had ended the old age and had given birth to the new one. It had opened up a "northwest passage" from the northwestern part of North America down through the middle of the great ice sheet and into what is now the Great Plains of the U.S. So, the comet was honored as the opener of gateways.

A supposed ancient Etruscan and Roman "superstition" about the crossing of the threshold and going through doorways now makes sense. They were not afraid of going through a gateway or door. They were remembering their high god, whom they credited with saving the world when he opened the gateway into the Americas. It was their way of giving thanks to their god.

We are only a little less "superstitious" about entering into the temples of our faiths, for when Catholics enter their Churches, they automatically reach for the holy water in order to make the sign of the cross with it and genuflect.

In fact, the rituals upon passing through the church gateway may have been first developed by the forebears of the Etruscans, handed down to the Romans, and then added to Catholic rituals performed in what had been pagan basilicas converted to Catholic use. We are not so far removed from those terrified Siberian hunters of long ago.

As precession of the equinoxes shifted their calendar dates against the seasons, conflicts developed between those wanting to honor the traditional calendar dates and those preferring to honor the summer solstice timing. The Romans tried to appease both camps. But the Etruscans and Romans were unable to keep their calendars in sync with the equinoxes. This led to a proliferation of feast days for the same event. So, Caesar recruited the Egyptian priest Sosigenes of Alexandria to "fix" the calendar. (Secundus (Pliny the Elder) 1991, Book 18, pp. 210-212) That, in turn, led to the disputes which stirred up widespread confusion and animosity and which contributed to Caesar's assassination. It was not until the 18th century that the British finally gave up on Caesar's "reformed" calendar and adopted the post-1582 Gregorian calendar we now use.

The long lags in calendar reform reveal how slowly such changes take place, and they show that calendars are like ancient fossils that can tell us about the past. In this case, the Etruscan-Roman calendar tells us that back at a time when the Summer Solstice occurred in their October, where our January now is, mankind was saved by a comet.

Every New Year's Eve, at midnight, we set off fireworks in the sky and watch a shining ball descend; we blow horns, hugging and kissing our loved ones, as if to say "we've survived" another disaster. We huddle together in the cold and wait for the crucial moment. And every so often, we are almost certain that it signifies The End of the World and that some cataclysm is going to overtake us. Not surprisingly, Rome kept a Festival of the Dead, on January first. (Gaster 1966, p. 45)

Some claim our calendar is off eleven days from the Mayan calendar. When Caesar's calendar was repaired by Pope Gregory in 1582, eleven days had to be adjusted for to correct the precession of the equinoxes in Caesar's calendar. So, we are now eleven days ahead on our calendar compared to the one in use at the beginning of 1582.

But, it is not our calendar that is in error. It was the old one of 1582 that was off by eleven days from the true length of the solar year. The 1582 dates had fallen out of sync for centuries; the error grew worse each year, reaching eleven days when Gregory corrected it. The Gregorian calendar simply restored the dates to what they should have been. (Duncan 1998, pp. 1-7)

So, December 21, 2012, is the Winter Solstice. It is not eleven-days short.

This brings us to the so-called "Mayan Calendar" that has stirred up so many people. First of all, it is not a calendar! It is technically a count-down of 1,872,000 days, or 5,125.366 of our years. (Gilbert and Cotterell 1996, p. 37)

That countdown is supposedly "amazingly accurate." But, it is not accurate at all. The total of 1,872,000 days was actually 5,200 years of 360 days, which would have been accurate only if the year had once been 360 days long for 5,200 consecutive years.

However, the late author Immanuel Velikovsky found evidence the year may have formerly been 360 days. (Velikovsky 1950, 1967, pp. 333-361) Unfortunately for the reputation of the Mayans, that was before Mayan civilization arose. So, the Mayans could not have directly observed such a 360-day year.

On the other hand, there are inscriptions on the stones of the much older Olmec civilization that utilized the same countdown. So, it was being kept before the Mayans. They simply inherited it from the Olmecs. (Richard 1999, p. 193) But, that leaves us wondering how the more primitive Olmecs came up with it. Fortunately, it does not matter who exactly had invented the countdown. It is the total count of the days that we need to address, not who gets credit for it.

The inscriptions allow us to count back to its starting point. The countdown began (expressed in our calendar) on the 11th or, some say, the 12th (or even 13th) of August, 3114 B.C. In both cases, they have it end on December 22, 2012, which would mean they have an error in their starting date or a miscalculation one day short of the required 1,872,000 days). (Gilbert and Cotterell 1996, pp. 2, 37) I should mention one source that says "most scholars" identify September 8, 3114 B.C. as the starting date and (inexplicably) December 23, 2012 as the ending date. (Richard 1999, p. 193) There must be an error, because the total count between the dates is 26 days short of 1,872,000 days. Other sources agree, however, on the December 23, 2012 end-date. (Duncan 1998, p. 19) All these sources show that the so-called Mayan end-date is not on the 2012 Winter Solstice.

The Mayan long-count ends perhaps a few days after the Winter Solstice (when earth's Northern Hemisphere is tilted directly away from the sun). But, contrary to some claims, this is not the day the sun is conjunct with the center of the Milky Way galaxy. That Milky Way/solar conjunction falls on December 19th, although in 2012, because it is a leap-year, the conjunction falls on December 18th. It will not be until about *A.D.* 2187 that the conjunction will coincide with the Winter Solstice. Therefore, the Winter Solstice of 2012 cannot be the moment of that Milky Way conjunction. And, even if it were, the Mayan end-date is at least a day or two (up to 28 days) after the Solstice. All who say something magical happens because of a Solstice/galactic/Mayan end-date conjunction are thoroughly deluded, because none of these conjunctions exist in 2012. None of the proposed Mayan end-dates is on the Solstice or the day of the galactic-solar crossing.

You will note that the cataclysm occurred about three Mayan Long Counts back in time (3 x 5125.366 = 15,376.1 years). While the Mayans knew this cataclysm happened, keep in mind that it had taken place 13,000 years before the Mayans ever carved an inscription. The preceding Olmec civilization had to tell them about the cataclysm.

And, you should recall that the Long Count is actually a simple reckoning of 5200 years of 360 days. The Mayans celebrated a festival, every 52 years. So, they kept (or hoped to) exactly 100 of these 52-year festivals in order to reckon the total of 5,200 years of 360-days. They also moved their capital every 400 years (13 x 400 = 5,200 years).

None of this makes any scientific sense today, nor did it when the Mayan Empire existed. They could not have based it on scientific observations of any kind. There are currently no such natural cycles as the 360-day, 52-year, or 400-year cycle. Nor did even Velikovsky make such a claim for the Mayan in their historical time-frame.

Fortunately, the Mayans had another calendar to rely on for practical purposes, while this one slowly fell into disuse. It was apparently more of an infrequent ceremonial artifice, not something they consulted on a daily basis.

You should remember also, that, regardless of which count or dates one uses, the actual day when the sun "eclipses" the black hole at the heart of the Milky Way galaxy will be (or was, if you are reading this afterwards) December 18ᵗʰ in 2012. The Solstice will be on December 21, not the 22ⁿᵈ as "Mayan calendar" writers Adrian Gilbert and Maurice Cotterell thought. The problem is that 2012 is a leap-year. So, there is a February 29ᵗʰ in it, making the Solstice a day sooner in December, on December 21, 2012.

We can conclude from all of this that our New Year's celebrations may have some lingering echo of the destruction of the *Nephilim*, but that determining dates is hopelessly muddled by all the calendar changes made by ancient (and modern) cultures.

The feast of St. John's Day on June 24ᵗʰ may preserve the same tradition kept in relation to the actual time of the Summer Solstice. Bonfires were lit on the Summer Solstice specifically in order to "drive away the lustful dragons" (a very apt description of the lustful *Nephilim*). (Gaster 1966, p. 36)

The day June 24 was apparently selected because it happened to be two days after the Solstice in Julius Caesar's new calendar, upon which ours is based.

As we have seen, February was followed by January in the old Roman calendar, but it had originally been in springtime, not in winter. Under Julius Caesar, 80 days were inserted into the calendar in 46 BC. So, the location of February (before 46 B.C.) was 80 days later than we now find it, that is, where we have April-May. (Duncan 1998, p. 33)

That is not all. The old Roman calendar had been a lunar calendar. So, the months moved around a few days more to make the 14ᵗʰ day of each lunar month the eve of the full moon, just as in the ancient Hebrew calendar.

This is not hard to understand when you recall that Rome said it had derived much of its religious practice from the Phoenicians, who were said to have founded the city. And, of course, the Phoenicians' King Hiram had helped build Solomon's Temple.

What all this means is that February once fell about where the month of Nisan, the month of Passover, falls in the Hebrew calendar, roughly April, as we noted. And that means that the following month, corresponding to our May, would have originally been named January.

Now I know this is probably getting very confusing, but it is important, as we shall discover. I have tried to proceed slowly, one point at a time to keep the complexity to a minimum. Just remember that as their calendars fell out of sync with the Solstice, some of the priests and the people objected that the date of the festivals were getting moved away from the true time of the events they commemorated.

So, some people kept fixed calendar dates to honor the event, while some calculated and kept the true seasonal timing, apparently about two days after the Summer Solstice (June 24, under our current reckoning).

And, when Caesar made his major adjustment to their calendars, a third group of people insisted on keeping the orphaned fixed calendar date's original time of the year.

So what day would this old orphaned date have been?

First, the old January 1st became the new March 1st (coming in "like a Lion") when the pre-Caesar priests had previously moved January from originally following February to coming before February. That meant March began to be right after February.

So, after they moved January before February, that left March 1 in the location where January 1 had been. As you can imagine, some people kept the old fixed feast on January 1, while others kept a new one on March 1, the time where January 1 had been.

But, when Julius Caesar moved everything another 80 days in 46 B.C., in a desperate effort to get the calendar back in sync with the seasons, he was moving his new May 1st into roughly the same time-frame in mid-spring where March 1 had just been, and where, before that, January 1 had once been.

Confused? Of course. And all of this hopelessly confused the Romans too, and it infuriated the priests in charge of the feast days. So, for this and many other reasons, they assassinated him.

Now, we can take a look at what May 1 really celebrated. You have probably heard of the May Pole. Well, it has a lot to do with our story. It symbolized the tilt of the axis, or the *axis mundi*, the idea of the North Pole's direction as the rotational axis in the heavens, in other words, the location of the North Star.

But of course, the ancients had by this time realized that the pole star changed as the seasons processed. So, if that point in the heavens were to wobble, as it probably did when the earth was jolted 15,000 years ago when the *Nephilim* were obliterated, then its symbol, the "May Pole," also wobbled. Hence, "dancing" around the "May Pole."

Among the Celts, May 1st was called "Beltane," which means, "The Fire of Bel (who was their god of the dead):" he was appeased by human sacrifice. It was said to be the first day of summer on the old calendar. All household fires were extinguished the night before and then relit from a single fire ignited by the priests. It was thought that on this same day (but different years) various peoples had all first migrated into Ireland (just as the extinction of the *Nephilim* marked the beginning of migrations into the New World). All this made May Day signify the idea of revolutionary change, the ideal day for socialists to choose for their parades. (Turner and Coulter 2001, pp. 95-96)

The first day of April, which is now where January 1st had once been, became "Fool's Day" and mocked the Fool, the firsts card of the Tarot deck, who is the vagabond Wanderer, Evander, whom the Bible calls Cain, also known as Osiris in Egypt and Romulus at Rome. His literal Egyptian name Asar (Osiris) appears in the Hebrew text of Genesis 4 and 6. several times: Asar/Osiris is called a leader of the *Nephilim*:

> ...they took... wives from all Asar chose... The *Gabarim* (Giants) were on earth in those days and also afterwards, when ben (a son of) Asar, of sons of the Elohim, came in to daughters of the (family) of Adam, and they bore... the giants of Asar from ancient-time, of the (family) of Enosh, the (son) of Shem. (Genesis 6:2, 4)

In the usual versions of this text, such as in the Authorized (King James) Version, the word Asar is translated by "whom" and "that" and "when" and "which were" and similar insignificant connective terms. The Authorized (King James)Version rendering of this passage is a valid reading, but it is not the only way it can be translated. Genesis was understood by most Israelites to have been composed by a man raised in Egypt, where the common name for Osiris was "Asar." And Osiris/Asar was acknowledged by the Egyptians to have been a giant and to have been sired by the "gods" and to have fathered "gods" himself. By the biblical reckoning, this would confirm that Asar/Osiris was considered one of the chief *Nephilim*.

So, we must allow for the possibility that this passage in the Bible was identifying one of the *Nephilim* by name: Asar, that is, Osiris (as the Greeks called him).

According to this alternate reading of the text, Asar himself chose daughters of the family of Adam to impregnate, and then afterwards, after the 15,000-year-ago cataclysm had wiped out the first wave of *Nephilim* in the Americas, a descendant of Asar again impregnated women of Adam's line, specifically descendants of the son of Shem named Enosh and the women of that family. This second set of illicit unions also had resulted in giants being born, who then corrupted the earth a second time, which led to the need for a second divine intervention, the Deluge in Noah's day.

There are a number of ancient pagan feasts that honor Asar/Osiris in one of his many guises, such as Dionysus, Bacchus, Evander, Lupercus, Romulus, *etc.* Thus, the rituals of May Day, or Beltane, like those of New Year's and St. John's Day and April Fool's Day, each reflect aspects of the *Nephilim* and the destruction of Asar's giants 15,000 years ago. All the complicated calendar changes have obscured the connections, but we can still detect the references to heavenly conflagrations, the opening of gateways to new migrations, the burning up of "lustful dragons," the abductions of Asar, and so on.

Of course, many practices were added to pagan holidays over the years that have no connection at all with the overarching celestial displays and horrifying events we have been examining. Mankind has long since lost sight of the original reasons for these festivals, even falling back into the worship of the *Nephilim* it had once dreaded.

Humanity twice faced annihilation at the hands of the *Nephilim*. In each case, God was said to have intervened. Noah's Flood is well-known. The earlier conflagration of the Americas has largely escaped notice. But there are some memorable accounts, such as this one from the descendants of the Siberian immigrants, the natives of the northwest: A "shooting star" (a comet) and a heavenly "fire-drill" (a primitive device used to start fires) were said to set the earth on fire:

> (Our ancestors) could see nothing but (tidal) waves of flames; rocks were burning; the earth was burning; everything was burning. Great rolls... of smoke were rising: Fire flew up toward the sky in flames in great sparks... The great fire was blazing, roaring all over the earth, burning rocks, earth, trees, trees, people, burning everything... Water rushed up... It rushed in like a crowd of rivers, covered the earth, and put out the fire as it rolled on toward the south... Water rose mountain high. (Velikovsky 1950, 1967, pp. 195-201)

The details in this account were not explored by Velikovsky because they did not fit his thesis that these legends described the plagues of the Exodus and the planet Venus. But these details are crucial: A comet (certainly not a typical brief "shooting star" or meteor) was seen in the sky for a long enough time prior to the conflagration that the people clearly remembered it as the cause.

True, it ended by acting like a shooting star and arcing toward the ground with a trail of fire as if it were a gigantic meteor. But we now know that the object left behind a scientific signature of diamond dust and ash covering a third of the earth that confirms it as a huge comet that exploded over North America and incinerated the continent. These details of the event are now an established fact.

However, it left the other side of the planet relatively unscathed, although the peoples of the Eastern Hemisphere suffered some losses from the violent winds and earthquakes that the comet's detonation triggered. As we have already discovered, their memories of the event were only dimly recalled; the Bible scarcely alludes to it. Most of their interest was directed at the fact that their women were no longer being abducted, at least for a time.

In other words, the Mideast recalled the cataclysm only as it affected their own world. The fire and flood were not witnessed by them. Jubilees merely states that the *Nephilim* and their "places" were all destroyed by "the Sword of the Lord." It says nothing about a fire or flood being involved, even though we know from the science that these cataclysms occurred.

Yet, this only begins to touch upon what native peoples recalled. Note the "waves of flames" that swept over the land. The tiny comet of 1908 in Siberia (also called "The Tunguska Explosion") generated a blast-wave that flash-burned and flattened forests for hundreds of square miles, like an atomic bomb. The huge comet 15,000 years ago would have been vastly more devastating.

Another confirming detail is the reference to everything burning, even all the people. This reveals that the native Americans who preserved the story knew that no one near the "shooting star" blast area survived it. They found nothing left alive in the region.

They also tell us that this huge sea of expanding fire was suddenly extinguished by a vast influx of waters flowing from the north like a mountain-high flood wall. These details exactly match what we know took place: The glacier in the north, up to three miles high, was instantly melted to the ground at its highest peak to create the corridor between the surviving parts of the flanks of the ice sheet. The melt-water then surged south toward the Gulf of Mexico in a mountain-high wall of water. It would have suddenly extinguished the fires, even as rock-melting super-hot as they appeared to be.

This tradition is so bizarre, so "impossible" in the normal course of events, that it can only be explained by the one event we know of when such a thing actually took place in North America: The suddenly melt-down of the heart of the ice-sheet 15,000 years ago when a huge comet exploded over North America and its white-hot fragments crashed into the glacier in a long string northwesterly from Illinois to Canada to Alaska to East Asia, melting walls of water that put out the sudden inferno and raced south to the Gulf.

Velikovsky found a number of dramatic references to this flash-burn exploding conflagration, but they came primarily from the Western Hemisphere and East Asia, not from his usual European and Mideast sources, where rains of fiery bitumen were more typical. (Velikovsky 1950, 1967, pp. 46-52, 64, 69-74) But, of course, we now can see why: It was because these great fires had erupted from a comet impact south of Siberia, in East Asia, and from the impacts along the ice corridor of North America. Therefore, it is from East Asia and the Americas that the stories of the world exploding with fire have come down to us. Keep in mind that the new Siberian immigrants peopled the whole Western Hemisphere within a few centuries of their arrival. (Pringle, *The First Americans* 2011, p. 41)

Velikovsky's ancient native American source had then continued, describing the appearance of this "shooting star" in terminology that can only have been derived from an actual ancient eyewitness account:

> (The monstrous comet descended) with a whistle... (that) made a terrible noise... (It) looked like an enormous bat with wings spread out (and that) grew larger until they could touch the sky on both sides. (Velikovsky 1950, 1967, p. 195)

The Wichita tribe adds more tantalizing details about "giants" being wiped out and monstrous (ice age?) animals dying. (Velikovsky 1950, 1967, pp. 195-196) They tell of all the animals fleeing before a wall of water that came from the north, and the four directions (the north pole) being disturbed (by something that must have jolted the planet); they emphasize the "re-peopling" of the land afterward. (Velikovsky 1950, 1967, pp. 195-196) All these things, we now realize, describe the same calamities that befell North America 15,000 years ago.

The "world ages" of ancient peoples did, however, usually include an age of man that was terminated by a great conflagration, just as they also remembered a global deluge of rain. The Greeks insisted there were great cycles of flood and fire, and the philosopher Heraclitus specifically gave the precise cycle between these global infernos as 10,800 years. (Velikovsky 1950, 1967, p. 46) This is about twice the Mayan cycle of 5,200 years. And like the Mayan countdown, Heraclitus' number was divisible by 360: 10,800 = 30 x 360.

Velikovsky found that reports of such cataclysmic cycles were especially detailed in the Americas, of which "all nations of this continent have preserved a more or less distinct memory." Of these, the Mayan inscriptions were among the most vivid, recounting "great catastrophes which, at intervals and repeatedly, convulsed the American continent." A repeating cycle was involved. The Mayans specifically cite one where the world was destroyed by Fire. The end of this age will also be accompanied by destruction, including fire. (Velikovsky 1950, 1967, pp. 47-52)

The rabbis too understood that several repeating ages of global destruction had occurred and would come again. Some rabbis explicitly cited long ages before Adam, with repeating epochs of 1,656 years separated from one another by destruction. This number is significant. It is the exact time-frame of the pre-flood world in Genesis, the era of the great *Nephilim* conflagration by an exploding comet.

Clearly, there is an awful secret hidden in these mysterious Cycles of Doom...

Chapter Bibliography

Duncan, David Ewing. 1998. *Calendar: Humanity's Epic Struggle to Determine a True and Accurate Year.* New York: Avon Books, Inc.

Gaster, Theodor Herzl. 1966. *Thespis: Ritual, myth, and drama in the ancient Near East.* New York: Harper Publishing.

Gilbert, Adrian, and Maurice Cotterell. 1996. *The Mayan Prophecies : Unlocking the Secrets of a Lost Civilization.* Rockport, Massachusetts: Element Books.

Gribbin, John, and Mary Gribbin. 1996. *Fire on Earth: Doomsday, Dinosaurs, and Humankind.* New York: St. Martin's Press.

Hogenboom, Melissa. 2016. BBC Earth: In Siberia in 1908, a huge explosion came out of nowhere. July 7. Accessed April 15, 2018. http://www.bbc.com/earth/story/20160706-in-siberia-in-1908-a-huge-explosion-came-out-of-nowhere. Additional reports can be found on https://earthsky.org/space/what-is-the-tunguska-explosion.

Mann, Charles C. 2011. "The Birth of Religion." *National Geographic Magazine,* June: pp. 34-59.

Pringle, Heather. 2011. "Texas Site Confirms Pre-Clovis Settlement of the Americas." *Science,* March 25: pp. 1512.

—. 2011. "The First Americans." *Scientific American,* November: pp. 36-45.

Quayle, Stephen. 2002. *Genesis 6 Giants.* Bozeman, Montana: End Time Thunder Publishers.

Richard, E. G. 1999. *Mapping Time: The Calandar and its History.* New York: Oxford University Press.

Secundus (Pliny the Elder), Gaius Plinius. 1991. *Natural History.* Translated by John F. Healy. London, England: Penquin Books.

Turner, Patricia, and Charles Russell Coulter. 2001. *Dictionary of Ancient Deities.* 1st. New York, New York: Oxford University Press.

Velikovsky, Immanuel. 1950, 1967. *Worlds in Collision.* New York: Dell Books.

Waters, Michael R., Steven L. Forman, Thomas A Jennings, Lee C. Nordt, Steven G. Driese, Joshua M. Feinberg, Joshua L. Keene, *et al.* 2011. "The Buttermilk Creek Complex and the Origins of Clovis at the Debra L. Friedkin Site, Texas." *Science,* March 25: 1599-1603.

Chapter 4

Cycles of Doom

The earth has been repeatedly battered by cataclysms caused by violent cosmic events. Two of these were said to have specifically targeted the Nephilim giants spawned by Asar. Just who was Asar? The name "Asar" is the original Egyptian name for "Osiris," which is actually the Greek version of his name. The "-is" is the masculine Greek grammatical ending for the word "Osir" (ah-s'r), which was how Greeks pronounced the Egyptian "Asar."

The ancient Egyptians insisted that Osiris-Asar was not native to the Nile valley. He and eight other members of his family (all later deified) had come from a land that was east of Egypt, according to the Pyramid Texts from the fifth and sixth dynasties in the Old Kingdom. These Pyramid Texts were inscribed on interior walls of pyramids constructed a few generations after the Giza complex where the Great Pyramid had been built. These texts give us enough data to retrace the wanderings of Osiris-Asar and his family. They lead us back to the Petra region in Jordan. Petra is about 50 miles (81 kilometers) south of Jerusalem.

Petra spoke a Semitic language. Asar's family had Semitic names, not Egyptian, a fact that even Zecharia Sitchin acknowledged. (Sitchin, *The Wars of Gods and Men* 1985, p. 38) (Sitchin, *The Twelfth Planet* 1985, p. 84) When "Asar" appears in Genesis, a book written by Israelites who had lived in Egypt, they knew Asar was the Egyptian Osiris. Moreover, they used "Asar" in the text of Genesis in a carefully-worded way that they had to realize would later be interpreted as a hidden reference to Osiris.

The Semitic word "AS'R" shows the roots of the name Asar. "AS'R" has varied meanings, depending on how it is vowel-pointed (Nikud). But, Nikuds were not used until *A.D.* 200. The original Genesis text could not specify just one meaning. All the grammatically-possible readings in context were legitimate (AS'R: *Strong's Exhaustive Concordance of the Bible*, Hebrew words nos. 833-839: "to be straight, level, right, happy, go forward, be honest, prosper, be blessed, be happy, go, guide, lead, relieve, who, which, what, where, that, when, how, because, in order that, Asher (person or tribe), happiness, going, a step, and (due to its upright growth) a cedar tree").

In addition, AS'R might also be read as "A-Sar," that is, "#1 Sar," or "chief ruler" (*cf. Strong's* # 8269) or "A-SAR," meaning, "first leaven (= first in pride)" or "chief remnant" or "chief survivor," or even, "elder flesh or kin" (*Strong's* Nos. 7603-7607).

The Egyptians reckoned Osiris-Asar as their "chief ruler." But he was one of a group of nine survivors who had wandered into Egypt at the time of a famine and taught agriculture to the people of the Nile valley, who made him their ruler in gratitude. We can see why the Egyptians might have considered Osiris-Asar to be a "guide" who "relieved" their famine and then "blessed" them with "prosperity" and "happiness"—all meanings that were associated with his name's Semitic origins, as shown above.

In Egypt, Asar was seen as a migratory farmer in a time of famine who had come from Petra with his sister-wife, Isis. This was her Greek name. Among the Egyptians, she was called AuSeT, the same Hebrew word used for Cain's "wife" in Genesis (Genesis 4:17).

A second century pagan who was an Isis devotee, Apuleius, listed many alternate names or titles for Isis in other ancient cultures: Ceres, Diana, Athena, Minerva, Neith, Hecate, and Proserpine (Persephone). (Walker 1983, p. 453-456) As Hecate and Persephone, Isis was Queen of the Dead, whom the Jews called Lilith. (Walker 1983, pp. 541-543) Thus, the Hebrew Lilith, as Isis, must accordingly be the wife of Osiris-Asar. Although she may have sought to seduce Adam, she was not his wife, as some authors have incorrectly assumed. (Ginzberg 1994, 1956, 1909, p. 34-35)

An alternate meaning of Asar as "first kin" could also indicate a "first-born" son, which the Egyptians said Asar was. Their Asar had a younger brother named "Seth."

Of course, we know that the Biblical Seth's eldest brother was Cain. The early Jewish-Christian work, *The Book of Adam and Eve,* said the first-born son, Cain, was born at a place south of the Jordan valley that is easily identifiable from the text as Petra. Exiled from the Petra region with his wife (AuSeT), Cain was a migratory farmer who was told the earth would "no longer yield its strength." A famine was coming. That this was a widely-understood meaning is confirmed by (*The Book of Jasher* 1924, 2:7).

The Egyptian Osiris-Asar was said to be from the first family of that age, closely paralleling the Biblical Cain. Both lived in the same generation of mankind. Both were born at or near Petra and both were forced to leave it. Both had a twin sister-wife known as "AuSeT." Both were first-born sons. Both were wandering migratory farmers. Both had a rival younger brother named Seth. Both featured tales of brotherly jealousy and revenge. And both had to contend with a famine not long after leaving Petra.

All we know of Cain matches Osiris-Asar, except that Egypt did not discuss Abel's murder. It is not surprising Egypt hid a murder committed by its benefactor. Instead, Egypt boasted of Osiris-Asar teaching the world to grow food. By contrast, they accused Isis of mating with Seth, although she claimed he had abducted her. (Sitchin, *The Wars of Gods and Men* 1985, p. 42) When she rescued Seth from Osiris' cloned son, Horus, the Egyptians felt free to say Horus had her beheaded. (Sitchin, *The Wars of Gods and Men* 1985, p. 45) But no ill was ascribed to Osiris/Asar.

The Egyptians also did not cite any infidelities by Osiris-Asar while he was away from Isis. But parallel accounts, like the Greek story of Dionysus, son of Zeus, are not shy about his sexual exploits. (Turner and Coulter 2001, pp. 152-153) He fathered many sons. Dionysus' father, Zeus (= Ra, father of Osiris), was associated with the rape and abduction of many women. (Turner and Coulter 2001, p. 522) That identifies Ra/Zeus as one of the Anunnaki or fallen angels. (Sitchin, *The Wars of Gods and Men* 1985, p. 124ff) So, Osiris-Asar and his wife Isis-AuSeT and their children (ultimately all of us) could all be properly classified as heirs of the Nephilim.

Previously, we saw that Genesis Chapter Six indicates Asar and his sons fathered *Nephilim* giants before the flood. The traditions of the rabbis were in no doubt that Satan (Ra) had fathered Cain (Asar) and that mankind had been corrupted by sex with fallen angels before the Flood. (Ginzberg 1994, 1956, 1909, pp. 54, 59, 63, 65, 68-70) Thus, the rabbis realized Cain-Asar and his wife Isis were giants. Noah alone was "perfect in his generations" (Genesis 6:8-12).

So why was Genesis carefully worded to hide allusions to Asar in its texts about Cain and the *Nephilim*? Why did it use secret double meanings to conceal negative statements about Asar and his sister-wife, that is, the true origins of Osiris and Isis?

The answer should be obvious: Genesis had originally been composed while the Israelites were slaves in Egypt (Genesis 50:20, *cf.* 50:15-21, 24-26). It contained intimate moments that only Joseph, while in Egypt, could have recorded (Genesis 42:24, 43:30-31). It ends with Joseph still buried and the Israelites still enslaved in Egypt (Genesis 50:24-26).

So, it was dangerous for Joseph or the Israelites to write unflattering statements about Asar and AuSeT while still at the "mercy" of the Egyptians. We must keep in mind that the later Greek names Osiris and Isis could not have been used in Genesis, which had been composed before there was a Greek language. Even Plato acknowledged this indirectly in his Timaeus dialogue, where it was noted that the Greek people, according to an Egyptian priest speaking around the time of Jeremiah, had "only recently" regained the use of writing, *i.e.* not long before c. 590 B.C.

The modern claim that Genesis was not an ancient text, but had supposedly been written during the Babylonian exile (after 586 B.C.) is refuted by numerous allusions to it that show it had originated long before the exile.[1]

Genesis was initially composed in Egypt, where it was dangerous to tell the truth about Osiris and Isis. Their history had to be hidden in case the book fell into Egyptian hands. It was the corruption and violence wrought by Egypt's *Nephilim* offspring that, we are told, had brought divine wrath upon the earth, descending from heaven with a destruction by fire or by deluge. They and their offspring were blamed for the repeated wrath of God.

There is a theological basis for linking these cataclysms with the *Nephilim*. God allows living beings freedom the way a gardener allows plants to flourish. But now and then, the gardener must trim back overgrowth and remove weeds or pests that threaten to destroy the garden. God knows "the end from the beginning." (Isaiah 46:10) Knowing the end makes God responsible for what happens if He does not intervene. We lament the loss of life from an intervention. Yet without it, "the end of all flesh" would occur. Ancient cultures said the giants had become voracious devourers of all living things and had to be destroyed to save humanity. From a theological perspective, God knew intervention was necessary in order to save the future world in which we now live. That is, none of us, or the rest of our world's living creatures, would be here now without these cataclysms.

---

[1] See Exodus 1:1-10, 3:6, 15-16, 6:3-4, 8, 14, 17:14, 19:3, 20:11, 23:10-12, 31:17, 33:1, Numbers 1:20, 26:5, 32:11, Deuteronomy 1:8, 11, 2:8-11, 20, 29, 3:11, 4:31-32, 9:27, Joshua 1:8, 8:30-35, 13:12, 17:1, Judges 4:11 ("Kenites" = "Cainites"), 21:3, Ruth 4:11-12, I Kings 6:23-29, I Chronicles 1:1-54, 2:1-4, 5:1-2, 29:18, Job 1:6-12, 2:1-7, 28:18, Psalms 8:3-6, 33:6, 104:1-32, 105:6, 23, 42, 110:4, 134:3, 136:5-9, 146:5-6, 148:1-7, Proverbs 3:18-20, 8:22-31, Ecclesiastes 12:1-2, 7, Isaiah 1:9-10, 3:9, 13:12, 34:8-11, 42:5, 45:12, 18, 63:16, 66:19, Jeremiah 4:23, 10:10-13, 50:40, Ezekiel 14:14, 20, 28:13-16. *etc.*

However, by definition, God sees the end from the very beginning. (*cf.* Isaiah 46:10, quoting Ecclesiastes 3:11 and Deuteronomy 11:12) His obligation to intervene exists from that point, from the creation. So, the interventions were created at the beginning. The changes in the sun, the exploding comets, the cycles of extinction and the plans for a remnant were all set in motion at the creation. Everything unfolds in order to allow as much freedom as possible, but without letting us self-destruct.

Genesis tells the same kind of stories about the pre-flood world as told by other ancient cultures. So, did all the ancients believe in the same exact cycles of doom?

> No... Although most counted either four or five ages or seven to nine of them, they had many varied theories concerning the length of each age. One said 10,800 years, others 3,600 years, and still others, 2,484. While short, these cycles were much longer than the Jews' supposed 1,656-year cycle. (Velikovsky 1950, 1967, pp. 54, 59, 63, 65, 68-70)

It is often assumed the Sumerians said the world was far older than the 6,000-year limit the Bible supposedly taught. We are told that Sumer had a long pre-flood history of 432,000 years. But this view is dependent upon interpreting the Sumerian "Sar" as always equal to 3,600 years. (Sitchin, *The Wars of Gods and Men* 1985, p. 79) There is reason to question this assumption.

Sumerians actually had two histories of the pre-flood world. One was their long 432,000-year Kings' List, but there was also a short 241,200-year Kings' List. (Pritchard 1969) (Oppenheim 1969, p. 265) (Sitchin, *The Twelfth Planet* 1985, pp. 248-250) (Sitchin, *The Wars of Gods and Men* 1985, pp. 76-77)

The reign of each king was given in a unit called a "Sar" that was believed to have been a time-period of 3,600 years. Thus, if the text said a king ruled for eight sars, then the translators assumed this meant he had ruled for 28,800 years. (Oppenheim 1969, p. 265)

The source of these dates was one man, Berossus, a Babylonian priest who wrote a history (now lost) promoting Babylon as the oldest culture in the world! Later Greek authors, such as Alexander Polyhistor, the philosopher Abydenus, and Apollodorus of Athens preserved Berossus' claim for the long Sumerian King's List and its ten kings before the Flood. According to these later writers, Berossus had claimed these Sumerian kings had ruled for 120 sars and that 120 sars equaled 432,000 years. (Sitchin, *The Twelfth Planet* 1985, p. 248) By the time the Sumerian tablets themselves were deciphered in the 19th century, scholars simply assumed Berossus' interpretation that a Sar equaled 3,600 years had been correct.

Scholars prefer dates that contradict Genesis. But the Sumerian meaning of "Sar" was a cycle of a star, planet, or constellation, but how long depended on the context. Berossus, writing 9,300 years after the Flood and 3,000 years after Sumer's tradition of it, was hardly able to fathom the intended meaning of their tablets. He made assumptions. But the total of 120 Sars (supposedly = 432,000 years) is a clue to the mystery.

In Berossus' day, just about the time his history was being written (*c.* 280 B.C.), the Greek LXX version of Genesis had just been published and placed in the then famous library of Alexandria, Egypt. It is not unlikely, given its direct relevance to his account of the Tower of Babel and other events he wrote about, that Berossus went to Alexandria to read it. Some scholars once wondered if Berossus were Jewish himself because of all his parallels with the Genesis account. (*The Catholic Encyclopedia* 1913, vol. II, P. 514a) The Septuagint Greek version of the Bible said that "man" (vs. "Adam" in the Hebrew version) was granted 120 "days" before the Flood (Genesis 6:3).

Berossus may have thought Sumerian texts about mankind's 120 sars before the flood were to be understood in light of the 120 pre-flood "days" granted "man" in Genesis.

There's more. Every 72 years, the sun will gradually precess one degree (move backward) around the zodiac. That makes the time of the equinoxes and solstices (and the seasons) drift backward against the stars. And, Berossus was not only a priest of Babylon, but also an astronomer, astrologer, and may perhaps have been world-famous for his ability to predict astronomical events. (*The Catholic Encyclopedia* 1913, p. 514-515) Berossus' astronomical skills could have led him to interpret a "sar" as some sort of precessional measurement. And, if we simply ask how many 72-year degrees there are in 120 of Berossus' sars, we discover it was exactly 6,000 (120 x 3,600 = 432,000 years; and 432,000/72 = 6,000 degrees of precession).

It is unlikely this 6,000 was accidental. The Greeks considered precession of the equinoxes was the "Great Year" of the sun. Each degree represented a Great Solar Day of 72 years. The Greeks worshiped the sun (Helios). Berossus might have seen great significance in an interpretation of 120 Sars that yielded a total of 6,000 degrees of precession or the sun passing exactly 200 zodiac "signs."

According to Berossus' interpretation, this 6,000 was the length of the pre-Flood solar age. There is today, as we all know, a wide-spread belief that the current age will end when 6,000 years' elapse. Of course, no such prophecy is in our Bibles. In fact, if one uses the Hebrew Old Testament (as the Authorized (King James) Version chronology does), the 6,000th year came roughly around A.D. 1850 (when the Adventists, Mormons, and other groups were expecting the "End"). But if one used the Greek Septuagint LXX version of Genesis, the 6,000th year came about *A.D.* 500-600 (due to two versions of the LXX chronology), about 1,500 years ago.

The original idea had been that six is the number of man and that God has therefore allotted six days to man, which are six of God's thousand-year divine "days" or 6,000 years. This idea goes back, it is thought, to the *Epistle of Barnabus*, which some date to the first century, but others reckon to be dated early in the second century. (Robinson 1976, p. 352) Other scholars relate this to the text, "In six days you shall do all your work, but on the seventh day you shall rest." (Exodus 20:8-11)

But, all such arguments must deal with the historical fact that both the Hebrew and the LXX chronologies of exactly 6,000 years have expired. The only way to cling to the 6,000-year idea is to mix various passages from each text and create a new version of the Bible that allows the 6,000th year to not yet have expired (a dubious method if one claims "divine authority" for one's view of prophecy). Whatever the length of this age may be, it is now more than 6,000 years if one uses either Biblical text as the starting point.

When the LXX 6,000-year count ran out in the sixth century, a new "prophet" arose, Mohammed, who insisted Christians and Jews were both confused about prophecy and the meaning of Scripture. He was able to persuade many Jews and Christians to convert to Islam, possibly in part because of the disillusionment felt by so many when the Greek LXX 6,000-year count had recently come and gone without the world age ending.

Biblical skepticism rose again when the 6,000-year count from the Hebrew text ended shortly after *A.D.* 1850, without the world ending, despite Adventists, Mormons, and many others insisting it would. Among believers, "date-setting" was condemned. But Jesus said only the exact day or hour is a mystery, not necessarily the year or even the specific season, which Jesus seems to identify as near summer (Mark 13:28-29):

> (At the End) the sun will be darkened and the moon not give its light... Learn the parable of the fig tree: When its branch is still tender and puts forth leaves, you know that summer is near. So you (also), in like manner, when you shall see these things come to pass (the fig leaves emerging just before summer and the sun and moon dark), you (should) know that it (summer and His return) is near, even at the doors (of summer). Truthfully, I say to you (that) this generation (that witnesses all these signs) will not pass away until all these things be finished. Heaven and earth shall pass away (for the appearance of the heavens will change and the conditions on the earth), but My words shall not pass away. But of that day and that hour knows no man, no not the angels who are in heaven, nor the Son,

but the Father (*cf.* Acts 1:7). Take heed, watch and pray, for you know not when the time is. It is like a man on a far journey who left his house in charge of servants... and commanded the porter to watch. Watch, therefore, for you know not when the Master of the house comes, at evening or at midnight or at sunrise or in the morning; lest, coming suddenly, he find you sleeping. And what I say to you, I say to all: Watch! (for the signs of the End) (Mark 13:24, 28-37)

Clearly, this is a command to watch for and study the predicted signs the age is about to end, not a warning to avoid thinking about, preparing for or anticipating the end! Jesus had just stated that the sun will be darkened, and the moon will not be giving its light at the time of the end. (Matthew 24:29, Mark 13:24, citing Isaiah 13:10 and Joel 2:31) In the apostles' time, the sunlight's shadow on a sundial was the primary method of determining the hour of the day. If the sun were to go dark, there was then no way to know the "hour" as they understood it. Even the ancient clepsydra (water clock) only measured a short length of time (the hourglass was not in existence at that time). As for the day of the month, it was counted from the light of the moon's visible phases. If the moon went dark, it would not have been possible to count the days, at least not as God reckons days in the Bible.

Before you assume that we could use a clock or calendar to find the time, Jesus had also said that "the days will be shortened," (*cf.* Matthew 24:22, Mark 13:20) and Revelation 8:12 understood that to mean a shortening of the length of each day by about one-third, to sixteen hours. Neither our clocks, timepieces, nor our calendars would be of any use in telling the time. We would have to create and distribute new ones, but that would be extraordinarily difficult during the conditions predicted for that "tribulation period." We will be literally "in the dark" about the exact time.

What all this means is that Jesus was simply warning that it will be physically impossible to know what day of the month it is or what time of day. But there is no condemnation for trying to discover the general timing. In fact, Jesus strongly urges us all to make every effort to watch for signs that it is coming. He is not against anticipating the time, but rather, Jesus only says it will not be possible to be as precise as the day or hour. It would be just as illogical to reject any anticipation of the timing of the end of the age as to reject any other topic because some people may be in error about it. We do not outlaw economics or science (or even sports!) because some people make predictions that fail.

The one thing Jesus does seem to indicate about the timing is that it will occur as summer is about to begin or shortly thereafter. Note the advice to "watch fig leaves" in order to determine the approximate arrival of the "doors" of summer (Mark 13:28-29).

What are the "doors of summer"?

The *Book of Enoch* (a book popular at that very time that emphasized signs of the end, see Jude 14-15) spoke of "doors" or "gates" of summer (Enoch 71-79). From Enoch, we know people in Jesus' time were aware there were several days either side of the summer solstice when the sun seemed to rise in the same spot on the horizon, as if stuck there. To the naked eye, the sun kept rising in the same place for several days, and these days were the "doors" of summer, days when the sun got "stuck" while trying to pass through narrow "gateways" in late June. (The same thing happened again around the winter solstice.)

In telling us to "watch" for signs of the End, Jesus drew special attention to these "doors" of summer, the several days that the sun seemed to be "stuck" either side of the actual day of the solstice. This period runs from about June 18th to about June 24th in a typical year. However, we should note that, with 16-hour days, it might run a few days longer, maybe ten days (*cf.* Rev 2:10).

Does this June 24th relate to all the evidence we examined previously that pagans were interested in the same time-frame? At least, they did when they were not haggling over which feast to observe as a memorial of the divine intervention that destroyed the *Nephilim* 15,000 years ago. The period of late June may also involve Pentecost...

Jesus said it would be like a man on a long journey in spring who is returning at the start of summer. To Jewish apostles, the man was going to miss keeping Pentecost, which usually comes a few weeks before summer begins. Recall that Jesus would ascend to heaven ten days before Pentecost, thereby "missing" the keeping Pentecost that very year.

Now the Bible has a specific law about this situation. Oddly enough, it does not make such a provision for any other time of year, only the Passover-Pentecost period.

Here's what God provided for in the Law of Moses: If a man had "touched a dead body" (which Jesus did in being crucified) or had gone away on a long journey and had missed the Passover, he could keep the feast as if it were in the second month, including all of the rules and regulations of the Passover cycle. (Numbers 9:2-14)

One of these rules of Passover was the counting of the Omer, 50 days until Pentecost (Numbers 9:3, Leviticus 23:4-21). That is, Pentecost was to occur 50 days after the Sabbath after Passover. (*cf.* Leviticus 23:16) Thus, if Passover was kept in the second month, the Pentecost would also be moved back several weeks as a result. And it happens that the time-frame in which this delayed Pentecost could then occur is right around the start of summer.

Pentecost was the latest day to release people from debt or free slaves in the year of Jubilee, which was based on the Passover cycle and the escape from Egypt. (Leviticus 25) These ideas relate to the Messianic Return and the Resurrection of the Saints and the termination of the persecution we are told will be occurring at that time (II Thessalonians 1:4-10, 2:1-12, Matthew 24:9, Revelation 13:7). Pentecost is therefore a credible time for the Second Coming.

We can see that Jesus was not picking some random imagery to use when describing the end of the age. The references to the doors of summer and to a man who had gone on a long journey were relevant to Jesus' personal timing and future return.

From the foregoing, we see that Jesus had a detailed view of the conditions that accompany the end of an age. By contrast, as we shall discover, the kind of information in most pagan accounts is vaguer. Indeed, when there is some verifiable information in pagan legends, it is often the same data already found in the Bible.

For example, pagan chronologies that described the pre-flood world contained the same ten kings or generations as found in Genesis and other ancient traditions (even the Chinese). The first king was often a form of Adam and the last was a Noah. The names were sometimes similar to Biblical names, and the flood stories also had many common elements with the Bible. From this we can infer that, whatever the claimed time-frame of the various lists, their solar "age" was the same one we read about in Genesis.

Yet, even their time-periods were not really that different, as our investigation of Berossus has already revealed. For example, Aristarchus of Samos, who lived not long after Berossus, said that every 2,484 years the world was destroyed twice, once by a vast conflagration of fire and then later by a global deluge of water. (Velikovsky 1950, 1967, p. 46) However, his 2,484 = 3 x 828, while the Biblical 1,656 = 2 x 828. So, there may have been a "leak" of the secret numbers of the Hebrew Genesis to the pagan Aristarchus.

Indeed, the dates in the Hebrew text had been a closely guarded secret. Jerome, 600 years after Aristarchus of Samos, said the meaning of Genesis was still kept secret by the rabbis. To learn how to read the text, Jerome met secretly with two rabbis late at night, so the rulers of the Jews would not find out. The two rabbis told him Genesis was to be explained only one-on-one, in a locked room without furniture or windows, and even then, only one verse at a time (*cf.* Jerome's preface to his translation of Ezekiel). These ancient rules had been in effect in the time of Berossus and Aristarchus. Therefore, the rules were in effect when the LXX Greek translation was done about that same time.

The *Septuagint Greek translation, the LXX,* was not the Jews' idea. The curiosity of the pagans of Alexandria, Egypt led to it. The ruler, Ptolemy II, had summoned the head of the Library of Alexandria, Demetrius Phalerum, and had asked him what was needed to "complete" their collection. The king's father was dying, and Ptolemy II wished to fulfill his father's dream of obtaining a Greek version of every important book. Demetrius said the main missing items were the books of the Jewish Bible. (*The Letter of Aristeas* 2010)

Unfortunately, pagans tried to monopolize books, outlawing copying or exporting (which led to great loss every time a library burned, or scrolls wore out from handling). Competition was fierce. Having the secret Jewish books would be a great coup. To entice the Jews, King Ptolemy II offered to buy back all the Jewish slaves captured in the recent wars. Not only that, he would pay for the recreation of the golden table of the Shewbread. But, he did insist the Jewish leaders come to Alexandria and do the translation there. The High Priest refused to come, but the Sanhedrin's remaining 71 members went. (*The Letter of Aristeas* 2010)

One suspects King Ptolemy II wanted the translators in a place where he could spy on them, or worse, threaten not to let them return home unless he was satisfied they had made a true translation. However, the first half of Genesis, the stories from Adam to Abraham, were a tithed portion to be read only by the priests and Levites (or those Jews in good standing whom they instructed privately), under penalty of death, like any "stealing" of tithes.

So, they could not do a true translation without violating the tithe. They had to fudge parts too sacred to reveal. Those who compared the Greek translation at Alexandria with the Hebrew text saw big differences, primarily in the tithed first half of the book of Genesis. Ezekiel, the tenth Hebrew Bible book (hence tithed), was also "badly" translated.

A scandal erupted... The Book of Ecclesiasticus or the Wisdom of Jesus ben Sira (not to be confused with the Biblical Book of Ecclesiastes) begins with an introduction in which the Greek translator admits that the two versions of the Bible were different. To quiet the protest, he tried to dismiss the differences as a "difficulty in rendering Hebrew into Greek." That, of course, was not the problem. The main difference was the Greek version added 50 or 100 years to the ages of patriarchs when they had their first son. Translating numbers like these is actually very easy to do correctly, and there were no such problems with numbers that did not affect the total lengths of chronologies. Clearly, the goal was to inflate the time from Adam to the Flood. They also inflated dates from the Flood to the Exodus by about 600 years.

Why would they want to do that?

First, they could not use the real numbers without violating the tithe. Second, they needed an alternative time-line that would seem credible to the people of Alexandria.

The Egyptians of Alexandria believed their first king, Menes, had united the nation in 3100 B.C. This same time-frame marked the beginning or revival of civilizations around the world, from China to India to Sumer to Egypt to Minoa to Stonehenge to Meso-America, all apparently recovering from a global disaster. If we multiply the dates in Genesis by seven (matching it with the date of the global flood 11,600 years ago), the Jews had been in captivity in Egypt in 3100 B.C., in the gap between the end of Genesis and the start of Exodus. So, the Jews had no account of these global events in *c.* 3100 B.C.

To match the Greek version of Genesis with the upheaval that had restarted global civilizations *c.* 3100 B.C., the Jewish translators decided to shift the Genesis Deluge back from the Hebrew date of *c.* 2500 B.C., to near 3100 B.C. That required adding a net total of about 600 years to the otherwise innocuous chronology of events after the Deluge.

Curiously, there are two versions of the Greek LXX translation of Genesis. They differ by exactly 100 years. If one places the Exodus *c.* 1485 B.C. (*cf.* 1 Kings 6:1), the two LXX Deluges are at 3,062 B.C. and 3,162 B.C. (Josephus 1960, p. 681, cf. pp. 678-708) No one knows why there are two versions. But when one averages the two dates, the result is 3112 B.C., two years shy of the starting date of 3114 B.C. for the "Mayan" calendar. (but see LXX Genesis 8: "...in the second year..."). This may be yet another coincidence, but as we shall discover, there are too many of these coincidences to explain them all away.

Aristarchus of Samos had heard the scandal about some sort of fakery in the LXX translation. He apparently made an investigation and discovered part of the Hebrew data, enough to generate his 2,484-year cycle, 1.5 times the Hebrew 1,656 period. He combined this with the Greek belief in a great conflagration, a fire he placed in the pre-Flood era. If his conflagration occurred well before the Deluge, it was well before 11,600 years ago. That would place it roughly at the time of the comet break-up 15,000 years ago.

Aristarchus was not the only one to attempt a solution to the question of how long between two ages. Heraclitus (who lived around 500 B.C.) reckoned the sun scorched the earth every 10,800 years. (Velikovsky 1950, 1967, p. 46) This is exactly three Sars of 3,600 years. But, it is also 150 degrees of precession of the equinoxes, equal to exactly five signs of the zodiac. This may relate to Revelation 9:1-10, an end-time prophecy about a five-month precessional sequence (= 10,800 years) from Sagittarius-Scorpio (Revelation 9:7) to Virgo-Leo (Revelation 9:8).

Another odd coincidence is that 10,800 years is roughly twice the "Mayan" cycle of 5,200 years of 360 days each. But we can be more precise than that.

If we take the Genesis 1,656 years and multiply by seven, we have a cycle of 11,592 years. This is based upon the fact that Genesis contains an artificially high frequency in its use of sevens, more than any other book of the Bible, including Revelation. Moreover, Joseph interprets Pharaoh's dream to mean that seven cattle equal seven years and that it represented a "fixed" prophetic time-pattern because God had repeated it (Genesis 41:25-32). We also have seen that Genesis contains secrets hidden in the text in case it fell into the hands of the Egyptians. The secret key to the true length of an age was surely one of the most precious secrets the Israelites possessed. Dividing it by seven and concealing the total in a complex genealogy of "ordinary" men was a simple way to hide the secret from the Egyptians, who were blinded by their pride in having such an "ancient" nation and by the glorious genealogies of their many "gods."

Did solar "scorching" come at the exact mid-point of the 11,592 years, or every 5,796 years? The global famine of Joseph did not occur then, but several centuries earlier: 5,400 years after the flood.

And that was also the start of the assumed Hebrew Genesis 6,000-year countdown. More to the point, twice this 5,400-year famine time is 10,800 years, the exact cycle of Heraclitus' solar scorching.

More coincidences? Consider the Egyptians. About the same time that Berossus was claiming great antiquity for his Babylonian civilization, there was a priest in Egypt, named Manetho, who was boasting of his own nation's great age. The text of Manetho is also now lost, but later writers have preserved portions of it, although they often revised it to fit their polemical needs. That has led to several different lists for Manetho's kings before the flood. In the following discussion, I have combined these to flesh out the descriptive portions, and have also supplied the Greek versions of the Egyptian names:

The oldest, Ptah (Greek writers called him Hephaestos), reigned 9,000 years and discovered the use of fire, clothing and pottery. These things imply a sudden colder climate had imposed itself while Hephaestos lived, requiring cooked food, a way to hold hot food and store it, and the making of warm clothing. Although the writers do not mention it, he was clearly trying to survive a sharp climate change. That suggests that an ice age had abruptly changed the way people then lived. Hence, we have an explanation for why Ptah is reckoned to have begun an "age" of mankind. He and his people had just survived some age-ending imposition of cold, apparently a sudden new ice-age. There is a hint the previous age had been hot and dry in the tradition that Sekhmet, goddess of the desert heat, had been his mother. (Turner and Coulter 2001, pp. 418-419) Since a similar transition may be the fate of our own generation, we ought to take note that this "high god" Ptah had been reduced to making fire, pottery, and warm clothing.

Next, came his son Ra (the Greeks assumed Ra was their Helios, although in the name of Amen-Ra, he could be identifiable with Zeus). (Turner and Coulter 2001, p. 522) Ra ruled for a 1,000 years. His identification with the sun suggests a warm spell, which we now know of as an inter-glacial warming like our own. They are said to last only about 12,000 years, more or less, and ours is already about 11,600 years old. Alternatively, the 1,000 years might be some sort of brief "millennial" interlude before full-scale glaciation. As we shall see, these initial Egyptian "gods" might have "reigned" in a pre-Adamic age.

Ra (the Greek sun-god Helios) was followed by Shu (unaccountably called Sosis, Savior, or Agatha-daemon, the Good Spirit, by Greek writers); he ruled 700* years. Shu was god of dryness, hence drought and famine (Turner, p. 428). Shu's sister/wife/queen was called Tefnut, "Lady of the cold, dark North," and the accounts of her said she mourned during a great famine. (Turner and Coulter 2001, pp. 456-457) Shu may be related to Tantalus, the king who was likewise the son of Zeus (Amen-Ra); Tantalus killed his own son Pelops and served him to the gods, possibly in a desperate attempt to evoke their sympathy for the plight of starving humanity. (Turner and Coulter 2001, p. 453) Pelops was quickly revived by the gods.

Geb (supposedly Cronos, father of Zeus) was Egypt's "earth father" ("Ge-Ab"). So, this links Geb with Adam. Geb "laid the cosmic egg" of which all humanity sprang; that "egg" was "Hawah" ("ova" or "egg"), the Hebrew "Eve" (mother of all living). Fertility had returned. Notice the repeating cycle as each new god ushers in a significant change in climate. The new age is then named after him, and he is given a length of life equal to his new age of climate. In Geb's case, his age is said to have lasted 500 years. (Turner and Coulter 2001, p. 189)

Geb's son, "legally" the first-born heir, was Osiris (Cain), who ruled with Isis, his sister-wife. This is another reason to identify Geb with Adam. We are told that Osiris and Isis rule for 450 years (keeping in mind that these periods are reconstructions of Manetho's lost history). Earlier we saw that Osiris and Isis entered Egypt during a time of drought and famine. So, the solar alternation in climate continued yet again.

According to Egypt, Osiris was abruptly overthrown by Seth (called Typhon by the Greeks). Seth supposedly violated the pattern that change in the kingship was to take place only when solar climate changed. In any case, Seth's rule is said to last 350 years (the *Turin Papyrus*, however, says only 200 years). The combined period of Osiris and Seth would be 800 years (or 650 years). The Egyptians felt that most of their combined period was a climatically bad one. (Manetho 1980, pp. 200 note 2-201)

After Seth was finally defeated, Horus, the seventh god-king, began to rule. His time was fondly recalled as signified by his symbol, the rising sun. Climate had improved for a time. But supposedly, only 300 years were left for him to rule before the Deluge.

There is uncertainty about what happened at this point. Thoth may have been the next ruler, eighth in the series. How long he ruled is debated. The Deluge may have already occurred, but we find that the unusually long period of 3,126 years was accorded him in the Turin Papyrus. (Manetho 1980, p. 3 note 1) All traditions of this post-flood period are rather murky and uncertain. Some versions suggest that the sun was not shining at all for a time, and that conditions were extremely harsh during the immediate post-flood era. So, we have yet another climatic reversal, continuing the pattern already established.

Cain's line in Genesis was a dynasty of seven generations before the flood. But Adam would make it eight. This makes sense when we remember that Osiris' family had also been Cain's family. So, the Egyptian lineage had seven or eight pre-flood deities.

However, it is clear that this Egyptian list goes back further than Adam, if we can identify Geb with him. Is there a pre-Adamic time-frame hidden in this list?

Actually, it is hardly hidden at all. Ignoring the odd 9,000-year time of Ptah, we have two distinct periods: 1,700 years before Geb-Adam and 1,600-years after that until the Deluge. These average 1,650 years, which is surprisingly close to the 1,656 years of the Bible. The Egyptian numbers were rounded off to the nearest 50 years of Manetho's chronology. Thus, the similarity to the Biblical data is striking.

The Biblical time-frame was not really different from the Egyptian record for the Geb-Adam to Deluge period. So, when the Israelites reduced their secret 11,592-year cyclical total by a factor of seven, it matched up well (if not perfectly) with this Egyptian tradition. The pre-Adamic 1,700-year Egyptian era then could add another 11,592 years to the chronology, putting the start of the age of Ra about 35,000 years ago, which is said by paleontologists to be when fully-modern man emerges in the fossil record.

If we also multiply the 9,000 years of Ptah by seven, we get 63,000 years, which (when added to the 35,000 years) gives us 98,000 years ago. That is when homo sapiens is said to first appear and master fire, just as Ptah is said to have first done.

Of course, this could be yet another coincidence. Or perhaps there was some truth in the ancient belief--reflected in the Book of Enoch--that fallen angels (the Nephilim) taught mankind the secret history of what had happened in those primordial epochs.

This brings us to Sumer, which boasted, according to Zecharia Sitchin, that it was in league with the *Nephilim* or *Anunnaki*, its gods. This Sumerian term "Anun-Naki" may mean the "Anun (heavenly) Nagas (reptilians)." Genesis calls Satan in the Garden the "Serpent" ("NaGaSh" = "Nagas") (*Strong's* #5175). These "Nagas" are found, under this same name (or very close to it) all over the ancient world. They were a notorious race of evil reptilian deities, who sometimes were said to live in underwater kingdoms. But this may be a confused reference to the *Nephilim* that drowned in the Deluge.

We can speculate that these Nagas or fallen angels had some sort of role in the form that the Sumerian Kings' Lists took. So, let's look at these lists.

The long list had ten kings, like the line from Adam to Noah, but the short list had eight kings, like Cain's lineage of seven or eight generations. This matches the Egyptian list of Ptah to Horus and its seven (or eighth) god-kings. The Eighth Egyptian god-king was Thoth. So, guess who was the Eighth king on the short Sumerian Kings' List? He was "Ubar-Tutu," which may be "Ubar-Thoth." (The Egyptian "Thoth" was "T'huti.") And just as the Egyptian list had Geb-Adam in the midst of their kings, the short Sumerian list had "Dumu-zi" in the fifth position. (Adam is also called "Ish" in the Hebrew text: Adam-Ish.)

The first king was "Alalim" (in Hebrew, this is "El-El-im" or "god of gods," that is, *Elohim*). This was followed by "Alalgar" ("beside Alalu"). Together the two ruled for 64,800 years, which equals 900 degrees of precession of the equinoxes. So, the Sumerian list began with 900 degrees of precession (= 30 zodiacal signs), while the Egyptian began with Ptah's 9,000 years (= 125 degrees). Another coincidence?

The remaining four god-kings on the short list had nearly identical names:

EN-MEM	-LU-ANNS	=	600 degrees
EN-MEN	-gal-ANNS	=	400 degrees
DUMU	-zi	=	500 degrees
EB-SIPA	-zi-ANNA	=	400 degrees
EN-MEN	-dur-ANNA	=	300 degrees

The total of this short list was 241,200 years, which sounds at first like a random number. It is not. It equals exactly 67 Sars of 3,600 years each. And it is 3,350 degrees of precession, or nine complete Solar Great Years of 360 degrees, plus 110 degrees.

Some versions have 18,000 years for Ubar-Tutu, the final king, while others have 18,600 years. (Pritchard 1969, p. 265) Could Ubar and Tutu represent two different kings?

What if there had been a ninth name on the list that got lost (as Thoth vanished from the end of some Egyptian lists)? The circumstance under which this is most likely to have happened would be if both had ruled an identical 18,000 years. Then the total for the whole list would have been 259,200 years (72 Sars) or 3,600 degrees (120 signs). This is exactly ten complete 25,920-year cycles of precession (10 Great Solar Years).

These nine kings could also then be clustered into four groups of 900 degrees each, or 40 times 90 degrees (a Great Solar season). This is yet another possible indication that this list originally had referred to ten periods of 360 degrees each.

Sumer had a very "symmetrical" chronology, whether in the long 432,000-year list (120 Sars = 6,000 degrees) or short 241,200/259,200-year list (67/72 Sars = 3350/3600 degrees =

120 precessional signs). Solar precession played the key role in both lists.

While all these priests and philosophers of the ancient pagan empires were competing with one another to make their nation the oldest on earth, the humble prophets of Israel, who knew the real secret, made no boast at all. Yet, everyone seems to have suspected that the Jews knew the real answers they were all seeking.

Indeed, Josephus, in quoting from the text of Berossus, which he had before him, comments that Berossus had published "books of astronomy (and), following the most ancient records (of his nation), gives a history of the deluge... and agrees with Moses' narration thereof... After which, he (Berossus) gives us a catalogue of the posterity of Noah and adds the years of their chronology..." (Josephus 1960, *Against Apion* 1:19)

Josephus had seen Berossus' original text with his own eyes and had studied it in detail, for he made great use of its text in his arguments for his book, Against Apion. It is clear from what Josephus says that he felt Berossus understood the Sumerian Kings Lists to be in some way mimicking the book of Genesis in detail.

Were both the Egyptians and the Sumerians emulating Genesis' chronology? Are their kings' lists more closely mirroring the Genesis account than we have found so far?

As we shall now see, the copying was more detailed than anyone has suspected. However, before we can show this, we must inspect Genesis itself more closely. Here are the times of rule from Adam to Noah, and next to them, the implied times of their climate "ages" named after them, plus the cumulative totals, both of which, we contend, were seven times longer than our English rendering assumes, as shown on the chart on the next page.

	World King, Time of Life, Time of Rule		Climate Age	Cumulative Climate Age
1.	Adam lived 930 years and ruled all 930 years.	x7	6,6510	6,510
2.	Seth rules for 110 years after Adam died.	x7	770	7,280
3.	Enosh ruled 98 years after Seth died.	x7	686	7,966
4.	Kenan ruled 95 years after Enosh died.	x7	665	8,631
5.	Mahalalel ruled 45 years after Kenan died.	x7	315	8,946
6.	Jared ruled 142 years after Mahalalel died.	x7	994	9,940
7.	Enoch never ruled; he was "taken" while Jared ruled.	x7	0	9,940
8.	Methuselah ruled 236+ years (until the Flood).	x7	1,652	11,592
9.	Lamech never ruled, but he lived 777 years.	x7	0	11,592
10.	Noah ruled after the Flood 350 years.	x7	350	14,042

Jared ruled exactly 10% of the time from Adam to the day of his own death, a "tithe" of the age to that point. During Jared's rule, Enoch, the Seventh patriarch, was "taken" and never ruled. Only seven kings (in red) actually ruled before the Flood. Methuselah, the Seventh king, ruled almost exactly one-seventh the pre-Flood era, until the day of the Flood.

It is details like these that have allowed some Bible scholars to dismiss the chronology as artificial, a man-made structure designed to incorporate a lot of arbitrary "sacred" numbers, mainly sevens and tens. They cannot believe any of this could actually have happened naturally; it is too non-random for chance.

But, that is the whole point. Either God Himself orchestrated these lives and the events in them, or else some author or editor modified the data in order to draw attention to these "sacred sevens" because they embodied some great secret that was worth all the man-hours it must have taken to create this list and its hidden design. (Or both are true.)

Only when we multiply all the numbers by seven can we see the true time when the rulerships took place, during the last ice-age. And only by adding up all those before the Flood do we arrive at the total length of the pre-Flood age: 11,592 years, the great secret the author was concealing from the Egyptians (and all those not privy to the portions of the text tithed only to the priests).

We can verify this. The comet that destroyed the *Nephilim* (for a time) and opened a corridor in the North American glacier has been scientifically dated to about 15,000 years ago. We also can calculate these 15,000-years back three Mayan calendar cycles (of 1,872,000 days), which brings us to November 16, 13,365 B.C.

If we are correct, not only will this day be near the summer solstice of that year, but it will also fall in the reign of the appropriate patriarch, and perhaps in a significant year in the Genesis history when multiplied by seven.

First, November 16, 13,365 B.C. was indeed the very day of the summer solstice, if we assume 15,376.098 years of solar precession (three Mayan cycles) during that period of time. (The solstice fell on November 16th if we ignore our modern calendar corrections.)

Second, this date falls within the rule of Enosh, the very person we previously had shown was actually named in the Genesis 6:4 account, where "and after that" refers to the 15,000-year ago comet explosion.

Third, the year of Enosh's rule, counting from the beginning of Adam, (the *Anno Mundi* or "year of the world") can be calculated. We must assume a date for the Exodus, namely, 1,485 B.C. This allows us to reckon a date for the beginning of Adam: 21,141 B.C. The comet explosion then falls in the 7,777th year after Adam, on the summer solstice. Even if we move the Exodus date around, we will be in the 7,700-year range for any Exodus date between 1,562 and 1,463 B.C. If the comet hit in the 7,777th year, then the elapsed time was just over 7776 years (7,776 = 6x6x6x6 = 6^5: that is, six to the fifth power).

We have support for a 21,141 B.C. time-frame from Sitchin's starting date for Ptah in the Egyptian kings' list. He makes it 17,900 years before Menes ruled Egypt. (Sitchin, *The Wars of Gods and Men* 1985, p. 126) That was at the start of the current Mayan calendar cycle in 3114 B.C. We must also add in 25 years (at least) of known rounding errors in the 17,900-years total, moving us back to 21039 B.C., roughly, for the first year of Ptah. (Manetho 1980, p. 5 note 6) That is 102 years after of our date for the start of Adam in the Genesis list. Since Egypt's list is really Cain's list, a slightly later date may be near his exile or the subsequent famine.

The various lists are turning out to be in remarkable agreement with one another, a fact that was not apparent until we began to dig deep into the matter. And that was, after all, precisely what the author of Genesis had intended. The secret has been so well hidden that even the priests themselves had forgotten its significance by the time of Jesus. He scolded them for losing track of the secret while they continued to restrict access to the tithed parts of Genesis to everyone else:

> Woe to you who are the doctors of the Law, for you have taken away the key of the knowledge (of the Torah's secret portions). You would not enter (into the secrets of the text) yourselves, and you prevented others (who wanted to rediscover them) from entering (into those secret things). (Luke 11:52)

The non-canonical *Book of Jasher* is mentioned in the Bible as a source text (see Joshua 10:13 and II Samuel 1:18), but there has been understandable skepticism about the Book of Jasher we now have, which may be an unrelated rabbinical text. We cannot resolve that issue, but it is worth noting that at least one passage in that text can now be verified:

> Seth called the name of his son Enosh, saying, because in that time the sons of men ("of enoshi" = Cain's line) began to multiply, and to afflict their souls by

transgressing and rebelling against God. And it was in the days of Enosh that the sons of men ("enoshi") continually rebelled and transgressed against God so that they magnified the wrath of YHWH against the sons of men ("enoshi")... and the sons of men ("enoshi") forsook YHWH all the days Enosh and of his descendants; and the wrath of YHWH was inflamed because of their (idolatrous) works and the abominations which they did in the land. And YHWH caused the waters of the River Gihon (or "waters exploding forth") to overwhelm them (the "enoshi"), and He destroyed them (with water) and consumed them (with fire), and He destroyed one-third of the earth...And in those days ...there was no food for the sons of men ("enoshi") and the famine was exceedingly great in those days...and the earth also became corrupt. And Enosh lived 90 years, and he begat Kenan. (Jasher 2:2-10)

The translator assumed the local River Gihon had flooded. Perhaps, but the rest of the description precisely fits the comet explosion that affected one-third of the planet and the ice melt-off that wiped away the Nephilim 15,000 years ago in the what may have been the 7,777th year of Adam, during the reign of Enosh as patriarch.

How could a local river overflowing have destroyed a third of the earth? "Gihon" means to "gush forth" explosively like a geyser. Therefore, "waters of gihon" must here be literally rendered as, "waters exploding forth" and must refer to water gushing southward in a tidal wave from the flash-melting of the North American glacier, an event we know occurred, and exactly at that time. The translator (in 1840) knew nothing of the comet 15,000 years ago or the melting of the glacier. He could only translate "gihon" as a reference to the Edenic river, not realizing it might refer to the flushing of the entire Missouri-Mississippi basin, from the Rockies to the Appalachians, on a single day.

After this comet event, there was a period of famine. This supports the evidence for the shaking of the earth by the comet impacts, which had wobbled the axis, disturbing climate and agriculture. Farmers always plant on the assumption of a repetition of past weather patterns. The comet events had changed everything. The expected weather did not occur... crops failed.

Why should this have affected the "sons of enoshi" more than the people of Enosh?

Cain's people, the *Nephilim enoshi*, were farmers, tied to the land. Enosh's people were herders and shepherds. The disturbed climate patterns affected Cain's people more than the line of Seth, because Seth's men were herders and could migrate.

So, we can see that, during the reign of Enosh, which was 15,000 years ago on the seven-fold Genesis time-line, the comets consumed the Nephilim enoshi with fire, and the ice of the huge glacier melted and washed them down the Mississippi basin into the Gulf of Mexico. The events so disturbed global climate that surviving enoshi farmers could not get their expected crops to grow, and they died of famine. With the end of these upheavals, a new era begins with the birth of Kenan. So here we have a specific reason for the climate-king pattern seen in the Egyptian King's List of Manetho.

But technically, the Egyptian list does not recognize any further Biblical rulers after the death of Seth. He is not succeeded by his son Enosh, but by Horus in the list of the Egyptians. Enosh is not a world-ruling patriarch as Adam and Seth had been. Horus, the heir of Cain, has claimed his birthright as the son of the first-born. As the rising sun in the east, Horus represents a time of prosperity and plenty.

But why should Horus be specifically identified with a sun rising in the east? Did not the sun always rise in the east?

Not according to the Egyptians. Immanuel Velikovsky gathered an impressive array of ancient traditions about repeated reversals of east and west. (Velikovsky 1950, 1967, p.118*ff*) He began by calling Herodotus to witness of the Egyptian priests, who told him that in the time since Egypt had become a kingdom (during the ice-age), "four times during this period (so they (the priests) told me (Herodotus)), the sun rose contrary to its custom: twice he rose where he now sets, and twice he set where he now rises" ((Velikovsky 1950, 1967, p.118*ff*), cf. Herodotus I:142).

Velikovsky noted that in the first century, Pomponius Mela, a Roman author, calculated the time from the start of Egyptian history to 13,725 B.C. That places it over 15,000 years ago, in the reign of Enosh, before the comet struck. He then describes how he read in the ancient Egyptian records about the four different sunrise orientations, including a changed course of the stars. (Velikovsky 1950, 1967, p. 119 note 7) Velikovsky realized that Mela was basing his comments upon an independent source from Herodotus, namely the written annals of Egypt, then still intact.

Could it be that the reason Horus was honored as the god of the eastern sunrise was because that was a new direction for the sunrise in the new age following the comet of Enosh's era? Once again, each new era was identified by a new ruler and "his" sun.

Ancient King's Lists are not merely attempts to get bragging rights for having the oldest kingdom. They are a code for periodic cycles of doom that repeatedly transformed the earth. Each king represented a new epoch of the planet in the aftermath of a great global destruction. The later lists were merely chronologies of rule by a single person. But the pre-flood lists were always a record of the great destructions and the world ages.

Let us look at the lists anew now, recognizing them as more like a precessional record of the history of the earth than of the personal lives of individuals; all the lists, even the one in Genesis, are divisible by the 72-year precessional cycle. That alone is evidence that their chronologies are really more about the times of the planet, rather than about the lives of those men who had to confront the cataclysms of the ages. (See chart on Page 73)

Some sources said Egypt's list totals 13,900 years, leaving 1,600 years missing, presumably those of the next god, T'huti. The Sumerian list starts out hopelessly behind by giving Alalim, the "Elohim" (God), only 400 units of rule. If they had given Alalim 9,000 units instead, like the Egyptians, their total would have been

	Sumer	Degrees of Precision	Genesis	KJV Seven-Fold Rule 21,141 B.C.		Egypt	Rounded-off Rule 21,039 B.C.	
				6,510 years	7,280		9,000 years	
						Ptah	(fire, clothing, pottery)	
1.	Alalim (Elohim)	400	Adam	930	7,280		900 x 10	
2.	Alalgar	500	Seth	770		Ra	1,000	
	Eridu	900	Eden		7,280			
3.	EnMenluAnna	600	Enosh	686	7,966			
4.	EnMengalAnna	400	Kenan	665	8,631			
		1,000					1,000	
5.	Missing king(s)	700	Mahalalel	315	8,946	Shu	700	
	Second Total	1,700	Three Kings	1,666			1,700	
6.	Dumuzi	500	Jared	994	9,940	Geb (Adam)	500	
7.	EnSipaziAnna	400	Enoch	-	-	Osirus (Cain	450	
	Third Total	900		994			950	
8.	EnMendurAnna	300	Methuselah	1,652	11,592	Seth	350	
9.	Ubar (son?)	250				Horus	300	
10.	Tutu	(250) ?	Noah	(2,450)		Thuti	(1,600)	
	Final Total	3,350			11,592		12,300	

11,950, almost exactly halfway between the Genesis (11,592) and Egyptian (12,300) totals.

From the general similarities in the groups, it appears that the 3,600-year Sar value of the Sumerian list was misunderstood by Berossus because he wanted to make Babylon the oldest civilization. Adjusting for this error, the Sar used was only 50 years. It seems the Egyptians regularly added 50 years to their totals to inflate Egypt's antiquity.

Close parallels among all the traditions indicate that a single underlying reality lay behind the chronologies. Genesis uniquely made no effort to exaggerate its numbers. Its totals were more specific and precise. It had no need to demote any of its kings. Thus, it appears Genesis provides the most accurate data about the Cycles of Doom.

Chapter Bibligraphy

Ginzberg, Louis. 1994, 1956, 1909. *Legends of the Jews*. Philadelphia, PA: Jewish Publication Society.

Josephus, Flavius. 1960. *The Complete Works*. Translated by William Whiston. Grand Rapids, Michigan: Kregel Publishing.

Manetho. 1980. *The Writings of Manetho*. Edited by G. F. Goold. Translated by W. G. Waddell. Cambridge, Massachusetts: Harvard University Press.

Oppenheim, A. Leo. 1969. *Babylonian and Assyrian Historical Texts*. 3rd. Edited by James B. Pritchard. Princeton, New Jersey: Princeton University Press.

Pritchard, James B. 1969. *Ancient Near East Texts*. Third. Princeton, New Jersey: Princeton University Press.

Robinson, John A.T. 1976. *Redating the New Testament*. Philadelphia, Pennsylvania: The Westminster Press.

Sitchin, Zecharia. 1985. *The Twelfth Planet*. New York: Avon Books.

—. 1985. *The Wars of Gods and Men*. New York: Avon Books.

The Book of Jasher. 1924. New York, New York: M.M. Noah and A.S. Gould.

The Catholic Encyclopedia. 1913. The Catholic Encyclopedia.

The Letter of Aristeas. 2010. Vol. I, in *The Old Testament Pseudepigrapha,* edited by James H Charlesworth, 33. Garden City, New York: Doubleday.

Turner, Patricia, and Charles Russell Coulter. 2001. *Dictionary of Ancient Dieties.* New York: Oxford University Press.

Velikovsky, Immanuel. 1950, 1967. *Worlds in Collision.* New York: Dell Books.

Walker, Barbara G. 1983. *The Woman's Encyclopedia of Myths and Secrets.* San Francisco, California: Harper and Row.

Chapter 5

Paths of Destruction

We have seen that the chronologies of Sumer, Egypt and Genesis, if we focus on their structural details, can all be largely reconciled with one another. They all describe the same solar age of about 12,000 years. Everyone understood that this previous world age ended with a solar-caused deluge. Their 12,000-year period was around half their estimate of 25,920 years for the complete 360-degree solar precessional cycle.

It was believed that precession of the equinoxes, by shifting the seasons around the celestial calendar, somehow triggered solar floods and famines. We saw that even modern scientists share some of this precessional-climate-change thinking. Genesis also uses the same precessional accounting of 72 years = one degree. Thus, Adam to the Flood (KJV) totaled 1,656 years = 23 x 72 years = 23 degrees of precession.

The ancients might have had only an approximation for how long precession took. The rate now is a bit less than 72 years, but it may be slowing. With such a moving target, one must pick a point in time to measure it. The Bible uses a 72-year rate, which was indeed correct at one time. Besides, in a world where an abacus was as high-tech as anyone could get in doing large calculations, the 72-year cycle was actually quite precise.

However, it is extremely important that the Hebrew text of Genesis uses a word for "year" that literally means "an enfolding" or "a repeating cycle of time." The reader discovers how long that time is from the context. Much of Hebrew is like that, requiring a study of the context in order to decipher what a word means in that specific case.

So, Genesis has a literal reading for the pre-flood age of 1,656 "cycles" of time. This can be literally read as 11,592 years if each cycle is a Sabbatical Cycle of 7 years. In fact, many of the great rabbis, such as Rashi, did indeed say Genesis went back 11,592 years, or seven times 1656. (Velikovsky, *Worlds in Collision* 1950, 1967, pp. 49-50) But, if that is how rabbis understood Genesis, then why has the whole non-rabbinical world adopted the short 1,656-year reading of the text?

It's no mystery. Everyone, even ordinary Jewish people, adopted the short "year" reading of the Septuagint (LXX) Greek text, whose translators were forbidden to disclose secrets in the initial tithed part of Genesis. One secret, of course, was that the chronology could be read as seven times longer. Thus, when they translated Genesis into Greek, they decided to use the short time-cycle, that is, a "year," rather than a Sabbatical cycle.

That Greek LXX translation dictates how our modern world now views Genesis. The LXX was begun *c.* 285 B.C., to please King Ptolemy II of Egypt, who "paid for" the translation of the Jewish Bible into Greek to complete his dying father's Alexandrian Library. Not only all gentiles, but nearly all the Jewish people themselves, for thousands of years, have derived our interpretations of Genesis from the translators' Greek LXX version. Accordingly, it is vital we know exactly how they came to use the short reading of the time-cycles, thereby hiding from the world the Genesis pre-flood ice-age history.

We have traditions of their translation from an Alexandrian Jew called Aristeas, writing within 75 years or less of the life of the king (by *c.* 170 B.C.). (J. H. Charlesworth 1985, pp. 8-9) We are told they swore a curse on anyone who edited or changed their Greek translation in any way. (*The Letter of Aristeas* 2010, pp. 33, 310-311) Note the detailed description of these events recorded by Aristeas (who claimed to have been an eye-witness):

> Demetrius of Phalerum (the chief librarian) endeavored through... negotiations (with other nations) to collect, if possible, all of the books in the world by purchase and by translation (including)...the books of the Law (Torah) of the Jews (which include Genesis) ...The king ordered a letter be written to the High Priest of the Jews... I (Aristeas) considered it an opportunity (to gain the release of) up to 100,000 (Jewish prisoners enslaved by the king's father, and so I) spoke these words to the king: "...The Laws (of Moses) have been established for all Jews. And so, it is our proposal not only to translate them, but also to interpret them (explain their secrets), but how can we justify such a mission while large numbers (of our people) are subjugated in your kingdom? But if because of your... generous soul, you release those... the God who gave them their Law (will) prosper your kingdom..." (So, the king) decreed that all Jewish people in slavery in the kingdom... be released, their owners to be compensated... When this was done, he ordered Demetrius to report on the copying of the books of the Jews. (Demetrius wrote:) "Scrolls of the Law of the Jews (the Torah) ...ought to be (placed in the library) in an accurate version, for this Law, as one should expect because of its divine nature, is... genuine... The whole army of historians have been reluctant to refer to (these books) and to the ancient men described in them (the Genesis history), because the examination of them is sacred and hallowed

(= restricted to priests) ...If you approve... (I shall request of) the High Priest... asking him to send men... skilled... concerning their Law, six from each tribe (= 72, as on the Sanhedrin), so that after the examination of the text, agreed on by the majority (of 72 men qualified to study that restricted sacred text), and after we achieve the (desired) accuracy in translation, we can make an exemplary (Greek) version (of the Torah books, including Genesis)." Eleazar, the High Priest (replied): "...Everything... we will carry out, even though it is not (our) custom (to make a) translation of the sacred Law... In the presence of the whole assembly, we selected elders... six from each tribe, whom we have sent with the law (to be kept) in their (exclusive) possession... (Please) order that once the translation of these books is complete, these men (and the law) be returned to us again in safety (not held hostage). The names of the men are as follows: (71 names are given, as would be expected if the High Priest sent the Sanhedrin but did not accompany them as its 72nd member)." (Aristeas said Eleazar) was very apprehensive (the king might retain such wise men at his court. Aristeas also said) "The leading priests of the Egyptians (which then included Manetho, had carried out) many careful investigations (of the Jews and their writings, as far as they were able) and declared us "Men of God." ...So they (the 71 tribal elders = the Sanhedrin) arrived (in Alexandria) ...with the fine parchments upon which the Law had been written in letters of gold (indicating it must date to Solomon's time, because he forbade any such golden Torah be done again, apparently in order that this golden copy become the standard reference text) ...They (after days of banqueting with the king) set about completing their individual assignments, reaching agreement among themselves (that is, conducting their work with the golden Torah scrolls in secrecy) by comparison of each of the versions. The (daily) result of their agreement then was made into an attractive copy (omitting crossed-out renderings and insertion marks) by Demetrius (who made the actual Greek copy that became the standard LXX reference copy in the library) ...When it was (all) completed, Demetrius read it aloud (at an assembly of the Jews of Alexandria). After the books were read, the priests stood up, with the (assent of the) elders from among the translators and... the (city's Jewish) community... and they (the priests) said, "Because this (Greek LXX) version has been done ... in all respects accurately (which they knew was not true, because its numbers were not the same as in the Hebrew Genesis), it is good that it remain exactly as it is and that there should be no revision," and they commanded that a curse should be placed, as was their custom (that is, the death penalty for violating tithed secrets), upon any person who would modify this (Greek LXX) version by any addition or change to any part of the written text, nor also any deletion (from it). (The king asked why no historian had referred to them before, and he was told) Because the Law was holy, being from God (hence, tithed as priestly), and indeed, some who tried (to reveal them) were punished by God due to their seeking to disclose the (secret) things of God to common man. (*The Letter of Aristeas* 2010, pp. 12-33)

Regardless of who Aristeas was, well prior to 170 B.C., Jews regarded parts of the Bible too sacred for "common man" (Deuteronomy 29:19-29). If God punished those who tried to disclose the secret portions of the Law, then would not God also punish these translators for disclosing the secrets? Clearly, the translators only pretended to do an accurate job, in order to get those 100,000 Jewish slaves released. Yet, no one today, neither scholar nor rabbi, contends that this Greek LXX version was accurate. They specifically cite its Genesis chronology as mistranslated. (Josephus 1960, pp. 680-684) (*Septuagint* 1977, pp. ii-vi) As we will show, even in Jesus' own day the authorities ridiculed those Jews who used that faulty translation as their Bible.

Early in the second century, the rabbis would sponsor another Greek translation (the Aquila version), which continued to hide the secret Genesis 11,592-year chronology. But, this version was for Jews who could read only Greek. The goal was to compete with Christian evangelists who were then using the Greek LXX version to promote prophecies in it that were worded in a way, they felt, was improperly pointing to Jesus. It was this aspect of the Greek LXX, not the falsified LXX chronology, that prompted the rabbis' Aquila version.

So, the rulers did not respect Greek Jews (although they accepted their tithes). During the fall feasts of *A.D.* 30, in the autumn of the year before the Crucifixion, a leader in the Sanhedrin (who would have known the LXX version was faulty), watching all the Greek-speaking foreign Jews crowding into Jerusalem for the feast (*cf.* Acts 2:7-11), said:

> But even this (non-Hebrew-speaking Jewish) crowd,
> those who do not know the Law (the actual Hebrew
> text of Genesis and the rest of Torah) are accursed
> (the curses for violating Torah in Deuteronomy
> 29:19-29)." (John 7:49)

Here they were, 300 years after the Septuagint LXX translation, accepting that any Jewish person who relied upon that bad translation and was ignorant of the original Hebrew was "cursed" by ignorance of the actual Hebrew Law of Moses. Moreover, because the first part of Genesis was tithed, most Jewish people were forbidden to read the original text, even if they could read Hebrew. The initial tithed portion was forbidden to anyone but the Levites, and the creation portion was strictly forbidden to anyone but the priests themselves or their initiates. "The secret things belong to the Lord our God (hence, to His priests), but the things revealed belong to us (= the rest of Israel) ...that we may do all the words of this Law (= "all these words which are written in this book," *i.e.*, the book of Deuteronomy: Deuteronomy 29:27)" (Deuteronomy 29:29). (Jeremias 1975, pp. 235-245)

The priests' wealth and power came from tithes. If they gave up their privileged status, they lost everything. Part of that privileged status as recipients of the tithes was their entitlement to these tithed secret portions of the Scripture. (Jeremias 1975, pp. 235-245) If they gave up one, they gave up the other. So, they dared not divulge to ordinary Jewish people the secrets of the Book of Genesis, for example, that it used a precessional cycle or that its text could be read as covering a time-frame at least seven times longer than non-priests realized.

Ancient cultures reckoned pre-flood chronology as less than half a precessional cycle. They limited that age to roughly a thousand years shy of 12,960 years (180 degrees of precession). The thinking seemed to be millennial. That is, there appeared to be roughly a thousand years "missing" between solar ages, as if the sun had gone "dark" for a while or had cast earth into prolonged gloom after the Deluge. In the book of Revelation, the sun is darkened, and the saints reside for a thousand years where "they need no... light of the sun" (Revelation 22:5). In the book of Zechariah, we are told, "it will be neither day nor night" (Zechariah 14:6-7). The Scandinavians and Teutons called it "the twilight of the gods."

What was this "twilight of the gods" in their legends? The phrase is one possible translation of a Norse word, Ragnarok, which might also mean, "dark rain of dust and ash." (Donnelly, *Ragnarok* 1883, 2004, p. 141) But another possibility might be to read it as "the fiery-light of the war-gods."

This revealed a great mystery. The Scandinavians have a unique (northern) perspective on solar conditions at the end of the age. They see the Aurora Borealis, the northern lights, on a regular basis. Long ago they noticed a link between the auroras and their weather. So, it was natural for Norsemen to conclude that the end of a great solar age would be accompanied by a far more severe change in the aurora.

The problem is that the aurora was often seen as a sinister-looking greenish light flickering like fire. And sometimes, in the dead of winter, they could see a "second" sun glowing in the extreme north. They assumed there was a mossy-green inferno at the pole, not sub-zero temperatures. They called this region "Muspell-heim" or a mossy-green realm of giants: "Moss-Bel-heim." Bal or Bel was a *Nephilim*-type giant, and he was identified with the sun, whom they called Bal-dur. (Turner and Coulter 2001, p. 90) So "Mossy-green Bel-heim" was a land where a greenish sun shone. They assumed that this land was inhabited by green giants who were made of green fire.

Norsemen were not alone in visualizing green giants. The father of the *Nephilim*, Osiris, the "night" (= northern) sun deity, was shown as green in Egypt. And, his giant analogues all over the world, usually dubbed "the Corn King" in honor of his role in agriculture, were also depicted as green or called "the Green Man" (especially in Britain). (Frazier 1890, 1981, pp. 305-307)

The Norse believed that the fiery-green giants of Muspell-heim, the realm of the glowing green sun, were drawing near their land when the green aurora brightened. Likewise, they believed the giants had retreated when the green auroral glow dimmed. At the end of the age, the fiery "green giants" were to gather the armies of Hell and ride south (from the Arctic) to make war on the heroes of the Norse country. (Turner and Coulter 2001, p. 444)

Thus, they expected the world to end when the light of the aurora grew bright and could be seen far to the south. The aurora grows brighter when solar storms increase in intensity. This is especially true during a coronal mass ejection, when part of the sun explodes out into space and collides with the earth's polar magnetic field, generating the massive electromagnetic discharge which we see as an aurora.

Fluctuations in solar activity over thousands of years do in fact reach extreme levels that can trigger global deluges and droughts. These are able to obliterate nearly all traces of human civilization and reduce humanity to a handful of survivors. The Norse have a front-row seat at the base of the I-Max auroral screen upon which the sun displays its coming destructions. The Norse have learned to interpret these haunting images as easily as the Chinese read tea-leaves. And in their old legends, they recall a terrifying time when "all of Hell literally broke loose." (Turner and Coulter 2001, p. 444)

They tell how the leader of the giants, Surtur, a "flame giant" bore aloft a great fiery sword, brighter than the sun, with whose falling embers he melted part of the Bilfrost of gigantic blocks of ice. (Turner and Coulter 2001, p. 444) This icy barrier was all that stood twixt this world and the flaming giants of Muspell-heim, "the abode of fire peopled by fiends." (Donnelly, *Ragnarok* 1883, 2004, p. 142) (Turner and Coulter 2001, p. 444) With Surtur comes Hell's legions and the Fenris Wolf, a gigantic demonic creature who "devours" the sun. Before and after them, as they ride through a hole torn in the heavens, are "flames of burning fire." (Turner and Coulter 2001, p. 444)

As Ignatius Donnelly realized in the 1880s, this description was both a prophetic warning and a history of a great comet breaking up and partially melting the northern ice. (Donnelly, *Ragnarok* 1883, 2004, p. 141*ff*)

We have already examined the story of the destruction of *Nephilim* giants in the Americas 15,000 years ago, when a great fiery comet exploded over the Midwest and partially melted the great North American glacier. The glacier had previously blocked the great cannibal giants of the Americas from moving north over the polar region where they might threaten Scandinavian lands. But when a corridor was flash-melted in the great ice-sheet, several giants might have escaped through it to the northlands. Since they were fleeing the fires, some could have been covered with the comet's burning naphtha (a flammable liquid hydrocarbon mixture).

The nightmarish sight of flaming cannibal giants riding through a flash-melted mountain of polar ice, with the sky glowing in brilliant colors, would not be easily forgotten. The flaming giants were not a silly myth. They were horribly real, and huge, some growing up to 32-feet tall, wielding giant swords that could slice a house in half...

> The legends insist a great battle with the giants followed, with the Norse heroes coming to the scene to defend the world from the flaming invaders. The giants were so large that even the greatest warriors could not withstand these gigantic fire-monsters. In the end, they all perished together. (Donnelly, *Ragnarok* 1883, 2004, p. 141*ff*)

It makes sense. The legends of each part of the world are describing the same event from their perspective. In America, they witnessed the exploding comet set the world ablaze, burning the giants they saw fleeing south, ahead of a mountain-high wall of glacial melt-water that swept them into the Gulf. The Norse saw the bright glow of the enflamed aurora just before a horde of burning giants fled northward through the newly-melted ice-corridor and invaded Scandinavia. The ice-age hunters in the Bering region west of Alaska saw comet explosions in Siberia and pursued terrified mammoth into Canada and down the corridor of ice into the devastated land. The books of Jubilees and Jasher tell how a third of the earth was destroyed by the "sword of the Lord," when God "consumed" the giants the first time, during the reign of the pre-flood patriarch Enosh (אֱנוֹשׁ translated Enos in the Authorized (King James) Version, see Genesis 5:9-11), and then washed the giant "sons of enoshy" away in an explosive flood. Those who survived perished in a great famine that resulted from the subsequent change in climate.

Modern scientists found the vast fan of diamond dust covering a third of the earth from a comet explosion 15,000 years ago. They documented the incinerated landscape, the flash-melting of the glacier, and the sudden opening of the ice corridor. They found that comets had also struck Alaska and Siberia and even Antarctica at that time. They found the actual artifacts of Siberian hunters who entered the Americas in the aftermath.

The implications are unsettling. How could such memories be preserved by all of these peoples? Was there not a global Deluge, out of which only eight people survived?

One of the great scientific issues raised by the Deluge is how all the mountains of the entire planet could have been simultaneously covered by water during the flood of Noah. But, there is a clue in the text where it says that the waters prevailed "fifteen cubits" above the mountains. (*cf.* Genesis 7:20)

Every mountain on earth is of different height. So, this can only refer to a single specific mountain where the depth of the water could actually be measured. We are told that the ark of Noah was 30 cubits high. (*cf.* Genesis 6:15) The clearance was scarcely half the height of the ark itself, that is, about where the fully-loaded ark would float in the waters.

If the ark had been built atop a mountain, the highest in the visible region, and then the flood rose enough to float the ark off the mountain, Noah would have concluded that the waters were at least fifteen cubits deep.

Building atop the mountain meant the ark would only be needed if the flood were mountain-high. Moreover, by the time the waters rose that high, there would be hardly any survivors left who might demand rescue. And by building it atop a mountain, few would realize it was intended to be a ship. In fact, The *Sibylline Oracles* tell us the three brides of Noah's sons (women of Cain's line: Genesis 4) thought the ark was a "house." (Fortner and Floyer 2011)

There is ancient confirmation that not all was covered by the Flood, in the priestly text of *Jubilees*, which scholars now date to before 140 B.C. (J. H. Charlesworth 1985, vol. 2, p. 44) The priest-author, who knew the Levitical viewpoint of his time, insisted that certain places were not covered by the Deluge. (Jubilees 4:24-26). While not conclusive by itself, the text shows the priests had long viewed the Deluge as targeting only those (vast) areas deemed wicked and infested with *Nephilim* corruption. (J. H. Charlesworth 1985, pp. 45, 63)

The first-hand account in Genesis is limited to what Noah could see from where he was. He saw his world water-covered, but perhaps certain parts of the globe were not. Genesis says, "the fountains of the great deep were broken up." And indeed, we know that a tectonic de-leveling event of this type was actually occurring at that same moment on the far side of the planet. It was sending a giant wall of water many hundreds of feet high, careening eastward across the Atlantic, building higher still as it funneled into the Mediterranean basin. It was aimed directly at the area where Noah was then living.

But more than tidal waves may have struck the Middle East land masses. Could the rising up of the Americas have been offset by the sinking down of continents in the eastern part of the world where Noah was? Note this prophecy about the coming end-time cataclysm:

> Prepare ye the way of the Lord; make his paths
> straight (or level, at the end of this age). Every valley
> shall be filled up, and every mountain and hill shall
> be brought low. (Luke 3:4-5)

In other words, as one area of the world rises, another sinks. Picture this image of the event as the ancient prophet Isaiah cried out for the Lord to come soon:

> Oh that Thou would rend the heavens and that
> Thou would come down, that the mountains might
> flow down (melt) at Thy presence... that the nations
> may tremble at Thy presence! (Isaiah 64:1-2)

Clearly, these are massive earth changes that were expected to accompany the end of the age. They involve a global reduction in the heights of all mountains, leveling them as if paving a highway across almost every place that is now a hill or mountain.

The end of the age will begin with a time when level paths of destruction will be paved across the world's mountain ranges. Jerusalem is to become the highest mountain on the earth. We will see eventually why this specific spot will become elevated when all else is reduced to a flattened (= humbled) status; there is a geological explanation for these events. For now, it is sufficient to note that, during the Deluge of Noah, while his region of the world was sinking under the waves, North America, which God had already cleansed of the Nephilim (the cause of the later Deluge), was rising up.

It makes sense. We know that a torrential "day and night of rain" had preceded the sinking of Atlantis. So, the Deluge was underway, and the rain was falling. Then suddenly there was this horrendous tectonic event, which yelped the whole ocean out of its banks. Gargantuan cliffs of water came screaming across the sea, sweeping up over the land, and lifting the ark of Noah off its mountain top.

This was only 11,600 years ago. It ought to have wiped out all the witnesses not on the ark. And it did in Noah's part of the world. He and those on the ark were the sole survivors.

But our understanding of the word "earth" in Genesis (and the rest of the Bible) is colored by our modern space-age view of "earth" as the whole planet. We assume that it always means the entire planet earth. But even we speak of "earth" as if it means "soil" or the visible "land" around our vicinity. For some reason, though, we reject this reading in the Bible. Yet why should the Bible speak of "the whole earth" or "all of the earth under heaven" and similar phrases, unless it needed to distinguish when it meant the entirety of the planet?

Accordingly, the idea that the entire planet was simultaneously under water comes down to the lone verse where we find the required phrase, "the waters were upon the face of the whole earth" (Genesis 8:9). It seems conclusive at first... until we read the context, in which we find that the ark had already come to rest upon the mountains of Ararat at the time in question. And this was months before "the whole earth" was still dealing with the muddy waters of the flood:

> And the waters retreated from the earth, going (away) and retreating, and the waters diminished at the end of the 150 days. And in the seventh month (at the end of 150 days of 30-day months from the day the flood began: Genesis 7:11), on the 17th day of the month, the ark rested on the mountains of Ararat. And the waters (from that time forward) were going

(away) and falling until the tenth month; in the tenth month, on the first day of the month, the tops of the (visible) mountains were seen (by Noah) ...and he sent out... the dove... but the dove found no rest for the sole of her foot, and she returned to him into the ark (where it was warm and dry and there was food to eat), for the waters were on the face of the whole earth (that is, soaking all the muddy ground Noah could now see plainly exposed, but still wet, so that the dove found no place nearby dry enough for it to not get its feet muddy, and there was nothing yet growing for it to eat, and so it naturally returned to the ark). (Genesis 8:3-9)

Whatever else might have been meant by the phrase "the whole earth" in this one place during the flood when it was used, it cannot mean that the mountain tops of the entire planet were then covered by water, for the text has just told us that the mountain tops in Noah's vicinity, those he could see, which is presumably all of them, were no longer covered. Therefore, the context is clearly describing what Noah can personally see from the ark, not what is happening on the far side of the planet.

That leaves us with the repeated description of the death of all the other living things "upon the face of the earth" (Genesis 7:23, *etc.*). If we are to be literal, that only includes those creatures that were out upon the visible "face" or the exposed surface of the open ground, not those in caves or other protected places during the flood. In fact, the natives of North America explicitly relate that they had survived the Deluge in caves and under the ground. North America is especially blessed with caves and rocky shelters.

We learn that an olive tree near Ararat had not only survived intact and alive but blossomed a few months later (Genesis 8:11). So, the plants did not all die.

Fish did not all die either, for Noah takes none on board the ark of which we are told, yet the world was still filled with thousands of fish species afterward.

So, then, what did die? "All those creatures in which was the breath of life (that were exposed) on the face of the earth died" (Genesis 7:23). Literally, every air-breathing and exposed creature died. This means that not only did the water drown them, but that it was also dangerous to breathe the air (volcanic gases?) or to be out under the downpour. Indeed, in the regions of the South Pacific, the Polynesian Deluge traditions speak of giant hailstones falling, and Revelation 16:21 warns of the same thing at the end of this age, as does Job 38:22-23.

Ignatius Donnelley, the author of *Atlantis, the Antediluvian World* (1882), even wrote an entire book, a 452-page sequel, to present the evidence about the rain of fire and gravel that he believed had accompanied the Deluge, arguing that it asphyxiated people and animals or pounded them to death. His second book outsold the first, possibly helped by the enormous explosion of the volcano Krakatoa, east of Java, in 1883. (Donnelly, *Ragnarok* 1883, 2004, p. 2)

During the Deluge, Donnelly said, it was unsafe to be outside and exposed. The ark was sealed shut, not only water-tight, but virtually air-tight as well (Genesis 7:16). But there may have been more breathable air in other parts of the world. (Donnelly, Ragnarok 1883, 2004, p. 1 map) Donnelly felt that the rain of fire and gravel had come from a massive comet, with the tail depositing a flammable and toxic cloud of petroleum gases for a couple thousand miles on either side of the Atlantic Ocean. (Donnelly, *Ragnarok* 1883, 2004, pp. 408-430)

Mr. Donnelly also made an interesting case for the Chicago fires of October 1871, being the result of a comet tail striking the earth. He documents the specific comet and the results. The most persuasive point is the deaths of people at a great distance from any fire, who died from inhaling toxic gases in open fields that never burned. He also cites the clouds of heavenly fire that fell out of the sky at the time. The area covered included four states, erupting simultaneously. (Donnelly, *Ragnarok* 1883, 2004, pp. 408-430)

The case for fiery comet encounters certainly does not depend upon the Chicago fire and the many other towns and cities that burned that same night. But, it is sobering to realize that such an event could have struck the United States that recently. In any case, seven years after Donnelly died, in June of 1908, a comet did explode over Siberia, and it blasted down a large forest and flash-burned it like an atomic bomb. (Called the "Tunguska Event," see (Lyne 1995))

More importantly, the events in 1871 and 1908 occurred in the same places where the comets of 15,000 and 65,000,000 years ago struck. The implication is that there may be some sort of astronomical-geologic connection we do not yet understand. That is, there may be a scientific reason why so many comets keep on following the same paths of destruction, from the Midwestern United States over the North Pole to the area of Siberia.

It is this curious fact that suggests that a comet may not have been the cause of massive deaths during the Deluge. Noah was not near this path of destruction. But if the sun were acting badly, as we believe, then radiation may have contributed to the deaths of those in exposed areas. In the *Book of Jasher*, we have further confirmation of bizarre solar activity at the time the Deluge began:

> And on that day (seven days before the Deluge
> began), the Lord caused the whole earth to shake,
> and the sun darkened, and the foundations of
> the world raged, and the whole earth was moved
> violently, and the lightning flashed and the thunder
> roared (as sometimes accompanies earthquakes),
> and all the fountains in the earth were broken up...
> (Jasher 3:11)

Most of this is standard fare for a great quake. But the sun darkening is not. That is expected, however, during a solar deluge event (Revelation 6:12, 16:10, Matthew 24:29, Acts 2:20, Isaiah 13:10, Ezekiel 32:7, *etc.*). While the rabbinical author of the *Book of Jasher* may indeed be drawing upon the rest of the Bible for the solar darkening idea, Genesis itself omits this part of the story, unless we read "breaking up of the fountains of the great Deep (יבשׁי *Tehom*)" as a reference to the "weeping" tears of the sun "Tammuz" or

"Tomas." This meant the sun was blamed for the rain, but it might also refer to a visible solar mass ejection that could be seen only if the sun went dark.

This brings us back to the Norse legends. They seem insistent that a substantial increase in auroral activity preceded the end of an age. The aurora is a direct result of solar coronal mass ejections. So, the sun gets super-active just before going dark. And we are told this directly and plainly in Revelation:

> And the fourth angel poured out its bowl upon the sun and power was given him to scorch men with fire. And men were scorched with great heat, and blasphemed the name of God who has power over these plagues... And the fifth angel poured out his bowl upon the throne of the Beast (whose kingdom at this point is the entire world) and his kingdom was full of darkness (indicating that the sun has now gone dark). And they gnawed their tongues for pain and blasphemed the God of heaven because of their pain and their sores (that would not heal = skin cancers: Revelation 16:2) and repented not of their deeds. (Revelation 16:8-11)

So, the sun first flares up, no doubt creating powerful auroral displays, and then it goes dark. The radiation is so intense that skin cancers become common. Everything is exactly as would be expected if the sun were to go through an extreme fluctuation.

Although the Norse legends speak of giants covered with fire riding through the partly melted glacier, an event we have identified with 15,000 years ago, that does not mean the same legends ignore the Deluge 11,600 years ago. In the Norse version, it has all been telescoped into a single tale, with overlapping stories.

For example, after Ragnarok destroys the world, there is a single surviving man named "Lif" (or Leif, a nickname for (O)laf, meaning the "aleph" or "alpha" man, the "first" and also "life"). He has hid inside their Tree of Life: Yggdrasil (Egg-worshipping-El = "The seed or fruit of those who seek or worship God"). The tree protects him, and after the old-world ends, a new one begins with Lif and his mate Lif-thrasir (eager for Life), from whom the new age is peopled. (Turner and Coulter 2001, pp. 291, 397) All this is their story of Adam and Eve and the Tree of Life. The Norse legend implies that a series of ages come to an end and are recreated in a similar manner, in a repeating cycle. (Turner and Coulter 2001, pp. 291, 397)

But, the Norse were not alone in believing in a pre-Adamic world. So, did the rabbis. They taught that there were six worlds created before Adam, and that each was destroyed. This world itself has undergone six upheavals, some by water and some by fire. (Velikovsky, *Worlds in Collision* 1950, 1967, p. 49-50) When the years of all these re-shapings and their destructions are added up, the totals are far greater than 6,000 years. The seven worlds were said to have each lasted 1,656 years, said rabbi Rashi, and accordingly, his total was 11,592 years. (Velikovsky, *Worlds in Collision* 1950, 1967, p. 50) The six preceding worlds would then total nearly 70,000 additional years. If Adam is reckoned to have lived *c.* 21,141 B.C., then the entire scope of the several creations began some 93,000 years ago. Other rabbis have had even larger totals, several times higher, rivaling the Babylonian totals given by Berossus. But, since they followed Berossus by several centuries, we may question whether their ideas were based upon an ancient tradition or were the result of simple competitiveness.

Likewise, the Norse legends have some familiar themes. The Luciferian personage Loki (= Lucky) is a shape-shifting giant who ate a heart and gave birth to three monsters, including Lilith-mimicking Hel, mistress of Hell. Loki was chained to a rock for causing the death of the sun (*Bal-Dar*). Poison dripped on his face, which his wife Siguna would catch in a bowl (*cf.* Prometheus chained to a rock for stealing heaven's fire for man; a vulture daily ate his liver, which regenerated until Hercules freed him). (Turner and Coulter 2001, pp. 390, 431)

Set free at the end of the age, Loki will pilot the ship Nagli-fari ("Naga's Ferry (to Hell)"), a vessel for souls of the angry dead

(*Nephilim* Nagas who died in previous world cleansings; "Naga" are Anunnaki, the fallen ones or *Nephilim*). The ferry Nagli-fari is kept moored on the border of *Nifl-heim* (*Nephilim*-land in the far south) and breaks free from *Nifl-heim* (Antarctica comet-strike 15,000 years ago?) and is swept north at the end of the age by a tidal wave of water unleashed by a gigantic, earth-girdling serpent. (Turner and Coulter 2001, pp. 252, 333) The 15,000-year ago comet known to have hit Antarctica may have swept a large ship full of dead giants into the north Atlantic, but this "ship that sails at the end of the age" story could have been mixed with Deluge legends as well.

Loki, like Cain, has two brothers, one of whom, Hoener (like Abel), is silent and exists only as "memory" as if dead in Hades. (Turner and Coulter 2001, p. 211) The other, Odin, is often identified with the Greek Mercury and shares some aspects of Zeus. (Turner and Coulter 2001, pp. 357-358) On the other hand, Loki, as the pilot of the Ferry to Hell, could be either Hades or Poseidon, the Greek sea god. Perhaps Loki must do his dead brother Hoener's job for him as part of his punishment for killing Baldar (Baal-Dar), the Norse sun god.

What did it mean that Loki caused the death of the sun?

We are told that, when Cain killed Abel, the ground would henceforth no longer be as productive as before. Moreover, an early Christian text said that a great earthquake had struck the planet. (Platt 2016, I: lxxix: 9-28) Both *Jubilees* and *Jasher* report that a famine occurred not long after Cain killed Abel. If something had happened to the sun and a great coronal mass ejection had hit the earth, it could explain both the quake and the altered solar output and fall in food production. Or it may be that the earth's axis was jolted out of position, which confused the growing seasons.

Perhaps the Norse thought Loki had caused agricultural output to decline because he disturbed solar activity when he caused the sun's "death." Prometheus, also chained to a rock, had "stolen" fire from the heavens, and in some versions of his story, from the sun itself. (Turner and Coulter 2001, p. 390) If the sun went dark or dim, it would explain why agriculture then failed and a famine occurred.

If our 11,592-year time-line for the pre-flood world is correct, then Adam would have begun *c.* 21,141 B.C. If Cain killed Abel not long after this, then there should have been a drought and famine that suddenly began 22,000 years ago, *c.* 20,000 B.C. There was. It was a sudden brutal era of drought, and it did not relent until 10,000 years ago. Arabia had been a pleasant savanna until then; afterward it became a cruel, arid sea of sand. (Clapp 1998, p. 275)

The 20,000 B.C. drought is about 10,800 years (= 3 x 3,600 years) before our flood date of 9549 B.C. This suggests some sort of cyclical link to the cataclysmic events we have been studying, ones which have their cause in heavenly cycles, not earthly ones. If the period were exactly 10,800 years, then the date this great drought began would have been *c.* 20,349 B.C. If we convert this date for the famine that began when Cain killed Abel into the more familiar 1,656-year chronology, Adam was then about 113 years old.

Genesis tells us that, when Adam was 130 years old, Adam and Eve had another child, Seth, as the replacement for Abel (Genesis 4:25; 5:3). That would have been seventeen years after Abel died. There is an early Christian tradition that Seth was born eight years after Abel's death. (Platt 2016, II:I:9, ii:3) It was felt that a few years elapsed after the murder before Adam and Eve dared to have another son. Abel's death was the first death they had ever witnessed, and even today, it is not uncommon for couples to be so devastated by such a loss that they wait a long time before having another child.

In any case, our estimated date of 20,349 B.C. is remarkably close, within nine years of the result produced by an early traditional date. We seem to be on the right track.

What happened in 20,349 B.C.? And, what was the 3,600-year cycle?

Zechariah Sitchin assumed that the 3,600-year Babylonian Sar cycle was the orbital period of a giant planet that he identified as Nibiru. (Sitchin, *The Twelfth Planet* 1985, p. 237*ff*) But, Sumerian scholar Dr. Michael Heiser, who has multiple doctorates in Semitic language studies (unlike Sitchin, who had none in the field), has said there are only nine mentions of Nibiru in the tablets and that all of them refer either to Jupiter or Mercury. Dr. Heiser, like nearly all other scholars with a knowledge of these matters, also rejects Sitchin's readings of the other words on the tablets, as well as his UFO-rocket ship interpretations of the Hebrew.

To be fair, Sitchin and his army of research assistants produced a ton of closely-reasoned books with massive amounts of ancient material and historical information in them. But, even though he provides a long list of extremely impressive scholarly works in the backs of his books in the bibliographies, there are only limited sources properly cited in his text, almost none of them specifying a specific page. I happen to have some of the basic original language sources in some of these areas, and despite having these, I could not find the origins of a number of Sitchin's claimed data and assertions.

For example, I have the Loeb Classical Library edition of *Manetho*. (Manetho 1980) It contains every single actual ancient Greek and Latin text that quotes or refers to Manetho's work on ancient Egyptian chronology... nothing is missing. But, I could not find a direct source for the exact chronological data that Sitchin presented in his book. (Sitchin, *The Wars of Gods and Men* 1985, p. 35) My guess is that he back-calculated it and then rounded off the numbers, the way I did, but he did not say so. Maybe, he used someone else's set of dates. Without footnotes in his books, I could not determine his sources, if any. Reading Sitchin is both fascinating and frustrating, and I have eleven of his books (so far).

He made claims about Nibiru colliding with another planet in the asteroid belt, and he said this was referred to in the Enuma Elish (a later Babylonian prayer to all the gods of the kingdom that was recited at the beginning of each year). So, I looked it up in *Pritchard's Ancient Near-Eastern Texts*, a work Sitchin lists in his extensive bibliography. It had been a standard scholarly reference on the topic for many years, although it has now been updated beyond the 1969 Princeton University edition Sitchin and I used. I mention this to point out that I checked the same edition Sitchin claimed he had access to.

I could not find anything that backed up his claim. But when you do this kind of work, dealing with hundreds of sources over the years, you inevitably make mistakes. Maybe, he was confused about where he had seen it. That sort of thing happens to all of us. Or maybe, he was doing one of his "creative" translations of an ancient text.

I can empathize with that too. Semitic language delights in presenting layers of meaning. Such alternate readings can be quite legitimate. Besides, no one really knows what some ancient authors had intended. In fact, it's possible that the ancient Babylonians who recited the *Enuma Elish* never fully understood it themselves. To some extent, all of us, even a professional Sumerologist like Michael Heiser, must make educated guesses about what these materials really mean. But, frankly, if I have to choose which person to rely upon in the meaning of Sumerian texts, Sitchin or Heiser: I think it wiser to go with Dr. Michael Heiser and identify Nibiru with Jupiter and/or Mercury.

Now, I suppose Sitchin would object that Nibiru had a period of 3,600 years, but that (supposedly) neither Mercury nor Jupiter has, or ever had, such a period.

But, Jupiter did have a 3,600-year period! We must remember, first, that the ancients did not have computers. So, they did approximations, sort of like the one Sitchin and I did to round off Manetho's chronology of pre-flood Egypt. It happens that Jupiter spends just about a year (361 days) in each sign of the zodiac. So, it takes Jupiter almost twelve of our current 365.242216-day years to circle the 360 degrees of the heavens. And 300 such Jupiter orbits = 3,600 years.

But, we can do much better than that. Jupiter spent exactly 361 days in each sign. If, as Velikovsky argued, there once had been a 360-day year, it would have precessed backwards at the rate of one day per year around the entire zodiac every 360 years. Meanwhile, Saturn would have made 12 revolutions during those same 360 years. And, Jupiter would precess exactly (to the day) ten times in 3,600 years.

Thus, in any culture that was using a 360-day year (and at one point or another nearly all ancient cultures made some use of a 360-day cycle), Jupiter (Nibiru) would be seen as a great 360-year "clock" whose motions could be used to measure a 3,600-year time-frame with amazing precision.

What about Mercury, the other "Nibiru" object?

Unlike Jupiter, which is so massive (more than all other planets and objects in the solar system combined, except for the sun itself), Mercury is rather tiny. The earth is 18 times the mass of Mercury. But, that makes Mercury uns table in its orbit. It can and does wander from its path. In fact, Mercury is the most unstable planet in the solar system, calculated to be drawn out of orbit and fly out past the earth in the far future.

Mercury may also have done that at some time in the past, and surprisingly, there is a great deal of evidence for this kind of errant behavior by Mercury.

For example, the entire original outer part of Mercury, including a good deal of its mantle, was torn away in a titanic impact with something enormous early in the history of the planet. Then, at a much later time, there was a second big collision, this time also with a huge object, but far less energetic than the first one. It produced an enormous impact crater upon the already devastated mega-collision surface. This second event, although smaller, was still a gargantuan impact that must have moved the planet out of whatever orbit it previously had.

Neither of these events was small. They could not have been a comet or asteroid impact. These gigantic collisions require Mercury to have hit another planet!

And, the two events must have happened right here in our vicinity, the inner solar system. Looking about this region for matching impact events, we find only two matches: A huge event on the earth in its infancy that formed our moon, and a more recent event on Mars that reshaped the whole surface of the red planet. Both of these are unsettling new discoveries by the astronomical community, which had been reluctant to embrace catastrophic cosmic events in our vicinity until July 16, 1994.

That was the day the Shoemaker-Levi comet, which had been broken into 21 bits by Jupiter's gravity, collided with the gas-giant, leaving behind 21 planet-sized dark blotches upon Jupiter's atmosphere. No one had expected any sign of the impacts. They assumed the tiny fragments of the comet would just be swallowed up by Jupiter's mushy frozen gases and vanish instantly from sight.

But, comets are not the innocent little puff-balls of ice and dust astronomers had wanted them to be. The fragments that hit Jupiter appear to have ignited in horrendous fiery blasts, leaving behind large patches of sooty debris the size of the earth. These were massive events. Had this happened to the earth instead, we would all be dead now.

Suddenly, the scientific community realized it had been working under some very significant false assumptions. Not only were impacts more likely than they had realized, but the devastation they could cause was vastly greater than anyone had imagined. No one dared say it out loud, but both Immanuel Velikovsky and Ignatius Donnelly had been right about how dangerous comets really were (as had William Whiston before them).

That 1994 event provoked rethinking about our moon's origin. None of the old ideas worked when simulated on computers. What if, some began to ask, there had been a big collision early in the earth's history? Could that explain the origin of the moon?

They tried several scenarios. At first, the effects of the computer runs were too big or too small. And then one almost worked. After a little tweaking, they realized it would work just fine if the object--as big as Mars--were to have hit the earth twice.

It was one thing to postulate a single event, but two was very controversial. Yet the data was a nearly perfect match. The requirement was to produce the kind of rocks found on the moon and the earth today. That's a big challenge, but the two-hit scenario worked. It meant that the first impact had to be a glancing blow, a passing jab, followed by a giant gouge of an impact, one big enough to excavate a glob of molten earth mantle material big enough to form the moon as we now find it.

There was just one lingering problem. What happened to the Mars-sized object?

The old tendency was to explain it away by saying it had merged with the earth or vaporized. The event would have been stupendously violent. The heat could have melted the impactor, but that does not mean it could not reform into a ball of molten matter and go on its merry way. After all, that was what they said the moon itself had done.

Had this Mars-sized object hit the earth like that, it would certainly have lost any semblance of its former size and surface. If it survived, all we might see today would be an object of maybe half its original mass and a thin crust of a surface over what would now be mostly iron-core material.

But, that is precisely what Mercury now is, a planet that has lost its outer surface and much of its mantle material to an utterly massive collision event.

Something else would have happened to that object that twice hit the earth. It would have lost a great deal of its orbital energy. It should have "fallen" toward the sun.

This could mean several things, depending upon the original orbit. After all, the fact that this planet had hit the earth at all implies one of the two planets had an errant orbit more like a comet than a planet. This is more likely to be the less massive planet, in this case, the Mars-sized impactor.

When the young earth was hit by Mercury (let's call it by its current name), Mercury must have been in a rather elongated orbit, one that may have threatened the planet Mars or Venus, as well as the earth. We did not just have bad luck to be the one that got hit. Our mass is greater than Mars or Venus; so, we had more attractive power to lure Mercury toward us. Our extra pull won out, and we got "smacked."

We were then a frozen ice-world, probably with a blanket of frozen gas as well. This cushioned the impact, allowing Mercury to survive. Mercury rebounded from what before the impact had been a frozen earth with one vast ocean of ice, but no continents.

What Mercury left behind was totally different. Shock-waves from the strike went through the earth and around the planet in concentric rings. These shock waves all came together in one spot on the direct opposite side of the planet. The earth's far side erupted in a great fountain of white-hot lava. Some of it may have been ejected out into space as a second large glob. After forming a second moon, it later impacted the first, and there is now evidence for this large lunar merger. (Barth 2012)

Meanwhile, the out-gushing lava spread across the globe like pancake batter pouring out onto a giant griddle. The flow continued for many years, eventually piling up the granite mass of which the continents are now formed. At first it formed one big pancake, a big fluffy lava platter, circular, and gradually oozing out to cover a quarter of the earth. That original continental mass is called Pangea by geologists.

The sun was not yet as warm as it is now. After several hundred million years, the earth cooled to the point that it refroze. For about three billion years or so, we remained more or less imprisoned in ice, a Sleeping Beauty waiting to be awakened by a princely gentleman. Too bad. We got beaten awake instead.

This time we were being clobbered by things a lot smaller than Mercury. But there were a lot of them. They melted the ice, unleashed the ocean currents, and began to fracture Pangea. This was a "mere" 600-million years ago.

Whatever it was, it battered all of the inner planets at the same time. Venus got slapped around. So, did Mercury and Mars. Even the moon got hit again and again. It seems that something had unleashed a violent barrage of debris all at once! That sounds like an explosion or some kind of destructive collision between what must have been two truly planet-sized objects, whose shattered remains could have generated the thousands of objects that pummeled the inner solar system 600-million years ago.

All these things–the wandering Mars-sized planet that became Mercury, the sudden massive battering 600-million years ago, the later super collision that happened to Mercury in more recent time--all these things are evidence that the solar system is not a nice peaceful place but has experienced some horrendously violent planetary collisions.

Did all such events only happen millions of years ago? Or have some happened during human history? Do planets collide or threaten each other now?

We mentioned that Mercury is highly unstable, the most unstable planet in the solar system today. Mars is second in that regard. Could it be that they are so unsteady in their orbits because they have recently disturbed one another?

Mars has one enormous impact crater that dwarfs all others. Its debris field is essentially the entire northern hemisphere of the planet. The crater is centered on its north pole. Whatever hit, it scored a bulls-eye. Or Mars is now 90 degrees out of its original orientation.

Indeed, the Mars pole actually is where the equator once was, and vice-versa. It may very well be that whatever hit Mars sent it tumbling until it reset its rotation as we now find it. In June of 2008, a cover article in the prestigious British science journal Nature argued that a Kuiper-Belt Object of at least 1,000-miles in diameter had hit the North Pole of Mars. Once again, we're told the impactor magically vanished after the attack. But did it?

Mercury also hit something very big. The crust is so thin that Mercury must have rung like a bell, its giant iron core leaving no place for a big impact to bury itself inside the planet. The impactor could only vaporize or bounce off.

Let's be serious. Are we really supposed to believe that three inner planets (the earth, Mars and Mercury) were hit by three independent giant objects that simply disappeared in big puffs of hot gas? Or, could it be that just one unstable, wandering object was responsible for all of these events?

Mercury is a perfect match for everything we have been discussing. Mercury was once Mars-sized. It hit something bigger than itself that stripped its outer covering away and left it to reform as the big iron ball it is today. That is what happened to the impactor that hit the earth to form our moon (and our continents).

Light-weight Mercury is prone to wander out of orbit. After hitting the earth, Mercury would have been redirected into a new orbit that returned it to our vicinity periodically, as if returning to the scene of the crime. This is, by the way, a very common and predictable phenomenon in orbital mechanics.

Mercury could not have had a circular orbit. The one it now has swings out some 43-million miles from the sun and then dives to nearly 27-million miles of it. Not even Pluto, which is sometimes inside the orbit of Neptune, has such an eccentric orbit. So, we can be sure that after it hit us, Mercury would have looped in and out of our vicinity, but not every year. These close encounters with earth would be spaced far apart, perhaps thousands of years apart, say, once every 3,600 years, for example.

We can prove this, if we can find two Mercury-earth encounters 3,600 years apart.

No problem... the first is easy. The *Gilgamesh Epic* from Sumer (*c.* 2,000 B.C.) includes the Deluge, and it shows that two planets were making close approaches to the earth as the Flood begins: Nergal (Mars) and Adad (= Thoth = Mercury). During its passages near the earth, Mercury often looked like a second moon in size and took on new names and titles in contrast to its usual "peaceful" aspect, but its attributes were those of Marduk (= Nibiru = Addu = Adad (Turner and Coulter 2001, p. 15)) when it threatened our planet:

> With the first light of dawn, a black cloud rose up from the horizon. Inside it, Adad (Mercury) thunders (bringing the Deluge of rain by showering the atmosphere with fine black dust that forms nuclei for condensation)... Erragal (Nergal (Mars): line 101 n205) rips out the dike-posts (that is, Mars' gravity triggers the tectonic uplift event in North America that sinks Atlantis and sends a mega-wave across the Atlantic into the Middle East)... Shock at Adad (Mercury), who turned all that had been light into utter-blackness, reaches to the heavens (that is, Mercury's black cloud alarmed the gods = the Anunnaki = the *Nephilim*). (*The Epic of Gilgamesh*, tablet xi, lines 96-106)

When we combine the effects of these two planetary close encounters with the other events occurring during the Deluge, it is easy to understand why God told Noah, "The end of all flesh has come before Me." (Genesis 6:13)

God did not allow that to happen, however. the whole point of the story is that God intervened to save both people and animals so that they did not entirely perish as a result of the cosmic upheavals that were coming upon the earth.

Using the seven-fold Sabbatical chronology, we have dated the Deluge to 9549 B.C., roughly matching Plato's Atlantis event, and coinciding with the global sea-level rise at the close of the most recent ice age. If we count 3,600 years (= one Sar) after this, we get 5949 B.C. Did any sudden cataclysms happen at that time?

Absolutely: The Black Sea was created when the Bosphorus broke through, and the basin beyond was flooded. The British Isles were created when their connection to the European mainland was severed. Volcanoes erupted. And the Tower of Babel fell, according to the seven-fold Sabbatical chronology. All this came 3,600 years after the flood.

Velikovsky, in his last work, published on the Internet posthumously, finally gave Mercury its due. He gathered evidence from around the world to show that Mercury had come by the earth and had caused the Tower of Babel events and other upheavals at that time. This was the explanation for Mercury being credited with feats of language and commerce and wisdom, he said. (Velikovsky, *The Dark Age of Greece* 2007)

So, now we have two consecutive Mercury-earth encounters 3,600 years apart. And if there were two, there must have been others. One of them, we now realize, came back in *c.* 20,349 B.C., when the earth shook after Cain killed Abel. At that same time, Mercury also caused a 12,000-year drought to begin in Arabia (among other things).

What other events can we trace to Mercury's close approaches?

If Mercury caused famines, then we might investigate the other famine we know about already, the one that struck 15,000 years ago, right at the end of Enosh's rule and just as Kenan took over (Jasher 2:7). How close was this to a Mercury flyby?

That previous Mercury encounter before the Deluge would have occurred in 13,149 B.C. By our seven-fold chronology, Enosh died in 13,161 B.C., twelve years earlier. That's almost a perfect match to Jasher's post-comet explosion famine 15,000 years ago.

If we go back another 3,600 years earlier, we arrive at c. 16,749 B.C. This is just 38 years after the 16,787 B.C. birth of Enoch in the seven-fold chronology. Recall that a new era of climate was named after its patriarch hero, and the Book of Enoch is unique in Hebrew literature for its description of deep snow and ice, which indeed was the climate in 16,747-to-16,749 B.C., at the peak of the last ice age.

We have one last event we can check, 3,600 years after the Tower of Babel/Black Sea event of 5,949 B.C. That would have been about 2,349 B.C.

Creationist geographer Donald W. Patten wrote a series of books attempting to precisely date biblical catastrophes caused by planetary close encounters. He calculated that one of these should have taken place in 2,376 B.C., which is just 27 years before our estimate for Mercury's expected 3,600-year close approach. (Patten 1988, p. 121) He had originally attributed many events to Mercury, but later switched to Mars as the exclusive cause, even when his ancient sources plainly identified Mercury. (Patten 1988, p. 223) But, in neither case could he (nor anyone else) find a hint of cataclysm in the years from 2,500-to-2,300 B.C. Not until the Exodus itself do any upheavals seem to have taken place.

This is no minor point. Generations of scientists, finding no trace of the Deluge at this epoch of history, which is the approximate time given in the KJV chronology, could only shrug and assume that the Genesis Flood was just a myth. Years ago, I personally talked with several of the top geologists working at the Lamont Doherty Geophysical Observatory, the premier institution in the field, asking them about any signs anywhere of a Deluge event at this time. Even in private, off-the-record conversations, they could not identify any. I did not then have the data to address the actual time of the Genesis Flood, around 9549 B.C., based upon the seven-fold Sabbatical cycle chronology.

Something had obviously gone amiss with the 3,600-year cycle of Mercury close approaches. Up to this point, we have identified six consecutive Mercury upheavals at 3600-year intervals between 20,349 B.C. and 5949 B.C. Yet, not even a significant volcanic event seems to have occurred c. 2349 B.C. (Knight and Lomas 2001, p. 58) The Nibiru "Sar" cycle of 3,600 years had ceased.

Mercury's orbit had been disturbed by something. Another object, big enough to shove little Mercury out of its comfortable niche, had come into our solar system...

Chapter Bibliography

Barth, Amy. 2012. "The Moon is Full of Surprises." *Discover Magazine*, March: pp. 26-28.

Charlesworth, James H. 1985. *The Old Testament Pseudepigrapha*. Vol. II. Garden City, New York: Doubleday Publishing.

Clapp, Nicholas. 1998. *The Road to Ubar: Finding the Atlantis of the Sands*. New York, New York: Houghton Mifflin Company.

Donnelly, Ignatius. 2013. *Atlantis, The Antediluvian World*. Seattle, WA: CreateSpace Independent Publishing Platform.

—. 1883, 2004. *The Destruction of Atlantis: Ragnarok, or the Age of Fire and Gravel*. Mineola, New York: Dover Publications, Inc.

—. 1974. *The Destruction of Atlantis: Ragnarok, the Age of Fire and Gravel*. Blauvelt, New York: Multimedia Publishing Corporation.

Fortner, Michael, and Sir John Floyer. 2011. *The Sibylline Oracles*. Lawton, Oklahoma: Great Plains Press.

Frazier, Sir George. 1890, 1981. *The Golden Bough: The Roots of Religion and Folklore*. New York, New York: Gramercy Books.

Jeremias, Joachim. 1975. *Jerusalem in the Time of Jesus*. Philadelphia: Fortress Press.

Josephus, Flavius. 1960. *The Complete Works*. Translated by William Whiston. Grand Rapids, Michigan: Kregel Publishing.

Knight, Christopher, and Robert Lomas. 2001. *Uriel's Machine: Uncovering the Secrets of Stonehenge, Noah's Flood and the Dawn of Civilization*. Gloucester, Massachuettts: Fair Winds Press.

Lyne, J. E. 1995. "Origin of the Tunguska Event." *Nature*, June: pp. 638-639.

Manetho. 1980. *The Writings of Manetho*. Edited by G. F. Goold. Translated by W. G. Waddell. Cambridge, Massachusetts: Harvard University Press.

Patten, Donald W. 1988. *Catastrophism and the Old Testament*. Seattle, Washington: Pacific Meridian Publishing.

Platt, Rutherford. 2016. *The Book of Adam and Eve*. Edited by Axioma. Vol. 1. 2 vols. New York, New York: Axioma Publishing.

Pritchard, James B. 1969. *Ancient Near East Texts*. Third. Princeton, New Jersey: Princeton University Press.

Sitchin, Zecharia. 1985. *The Twelfth Planet*. New York: Avon Books.

—. 1985. *The Wars of Gods and Men*. New York: Avon Books .

The Letter of Aristeas. 2010. Vol. I, in *The Old Testament Pseudepigrapha*, edited by James H Charlesworth, 33. Garden City, New York: Doubleday.

1977. *The Septuagint Version of the Old Testament*. Grand Rapids, Michigan: Zondervan Press.

Turner, Patricia, and Charles Russell Coulter. 2001. *Dictionary of Ancient Deities*. 1st. New York, New York: Oxford University Press.

Velikovsky, Immanuel. 2007. "The Dark Age of Greece." *The Velilovsky Archive*. August 29. Accessed April 30, 2018. https://www.varchive.org/dag/index.htm.

—. 1950, 1967. *Worlds in Collision*. New York: Dell Books.

Chapter 6

The Typhon Mystery

Many people wrongly assume the "literal" Bible says the world was created 6,000 years ago. Yet, for millennia, the rabbis have believed in multiple long ages and have based their belief in the Hebrew text of Genesis itself. In fact, the rabbinical age for the world was generally longer than that of pagan world. Only a few pagans felt the world was far older.

Besides the already-discussed 432,000-year chronologies of Berossus and the Sumerians, there was an even greater claim in the Hindu version of the pre-flood world. The Hindu story had fourteen patriarchs or "Manus" ruling an age of 4,320,000 years. This is the same number used by Berossus but multiplied by ten. This Hindu number was also exactly ten times 6,000 degrees of precession of the equinoxes. A Greek tradition likewise called the spirits of the pre-flood dead, the "Manes." (Frazer 1964, p. 179)

Each Hindu Manu gave his name to the age he ushered in. (Turner and Coulter 2001, p. 307) In Semitic letters, Manu could be written as "Mem-Nun" or the familiar "Memnon," and in "Agamemnon," who were Greek rulers of c. 3,100 B.C. Other Manus were: Menes (Egyptian) and Minos (Minoan or Cretan), two conquerors who also date to c. 3,100 B.C. (but who might be the same person), implying that a new "age" began c. 3,100 B.C. India, China, northern Europe, the Americas and Sumer also "restart" their civilizations at this time. (Knight and Lomas 2001) So, prior to 3,100 B.C., the earth must have had an age-ending global cataclysm. Jewish leaders in 285 B.C. were so confident that a deluge date of c. 3,100 B.C. would be fully accepted by Greeks and Egyptians, they inflated their *Septuagint LXX* chronology in Genesis in order to force its Flood date to artificially match 3,100 B.C.

The Greeks and Egyptians said the pre-flood world had originally been ruled by ten twin brothers. Likewise, the Chinese, the Iranians (Persians), the Brahmins, the Germans, and the Arabs knew of ten such primordial rulers. (Donnelly 1882, 2013, pp. 13-14, 27)We saw that Adam in Genesis had ten heirs, but two lines of seven kings each, one from Cain and one from his brother Seth. Thus, the fourteen Manus of India were really the same fourteen kings we encounter in Genesis, which in turn are related to the ten patriarchs of Greek, Egyptian, Chinese, Iranian, Brahmin, German, and Arabic tradition.

Names may change from culture to culture, but the overall story stays the same. For example, each Hindu age ends in a Great Flood (Turner and Coulter 2001, p. 307-308). We are in the seventh great age, as the rabbis also said, along with the Mayans, the Brahmins, the Buddhists, the Persians, the Etruscans, the Sibyls and many others. (Velikovsky 1950, 1967, pp. 46-52) The eventual flood survivor is told to build a sea-worthy ship; he spends a long time on board, lands on a high mountain, and offers an edible sacrifice after the waters recede. He also re-populates the world with animals. However, the silly Hindu version veers from Genesis by adding a talking fish and animal-generation via shape-shifting incest. (Velikovsky 1950, 1967, pp. 46-52)

It is true that the Hindu system allowed for billions of years, but only because it was an open-ended cycle that could be extended forever. On the other hand, as we saw, the Sumerian ages had also been puffed up to make their nation seem older than it was.

So, let's ignore everyone's inflated chronologies. What we see is that some Jewish rabbis had preserved an ancient secret Israelite tradition of a series of 11,592-year ages, nicely matching the ice-age cycles now fully verified by modern scientific investigations. While everyone was publicly making boastful fictitious claims in their histories (even the *Septuagint* translators), often inventing silly myths, an ancient Israelite line of wise men and their trusted rabbinical heirs were secretly studying the true ages and preserving an accurate record of ordinary human beings surviving actual geologic upheavals.

One thing is certain: Everyone in the old world, pagan or not, believed there had been multiple ages of the earth, each one ending with some kind of cataclysm.

As we have already discovered, almost everyone knew of a world age that ended in a conflagration of cosmic fire that fell from the sky. We now know of at least one event that perfectly fits this description, and it happened about 15,000 years ago, around 13,365 B.C. And, of course, everyone knew of the *c.* 9,600 B.C. global deluge of torrential rain that ultimately ended the age that came after that great destruction by cosmic fire.

Everyone also knew of other age-ending calamities: Some ended with global super-quakes that reshaped the mountains, changed the course of rivers, and tilted the axis of the earth. Some ended in mega-volcanism and titanic wind. (Velikovsky 1950, 1967, pp. 47-48) Still others were terminated by all the oceans rising up and washing civilization off the earth.

Curiously, only a handful mentioned ages that ended when ice or deluges of hailstones buried continents. Besides the Jewish people and Christians (prophetically), the Icelanders and the Polynesians were among the few who described enough ice falling from the sky to bring an age to an end. It seems the most recent events in human memory were those involving deluges of rain and tidal waves of water, not ice.

Yet, we now realize that the dominant feature of climate change over the previous geologic eons has been repeated cycling of ice ages when global temperatures plunged, sea level fell, and vast sheets of ice abruptly accumulated on the land. Jews and a few others knew of such events, but the vaunted Sumerians, Egyptians, and peoples of India recalled little or nothing of them. These cultures had no record of cyclical deluges of ice.

But, let us not be too hard on the Egyptians, for they seemed to know what was causing at least some of the cycles. Listen carefully to what the aged Egyptian priest-historian and scientist told the Greek Statesman Solon in c. 580 B.C., when Solon visited the already 8,000-year old city of Sais on a branch of the Nile, just upriver from where Alexandria would be built about 250 years after their conversation:

> You Greeks are but children... There is no old wisdom handed down among you by ancient tradition (as Israel had), nor any science hoary with age (as Egypt had). And, I will tell you the reason: there have been and there will be again many destructions of mankind due to multiple causes... You have a story that Phaethon, son of Helios (the sun), yoked the horses of his father's chariot, but unable to drive them on the path (the sun takes through the sky), burnt up everything then upon the earth... Now, this tale has the form of a myth, but in reality, it refers to a Declination (a declining down) of those cosmic bodies that orbit near the earth through the heavens (close enough to disturb the earth's rotation and the sun's apparent path), and a Great Conflagration of things upon the earth, that has been recurring over long intervals (cycles) of time. (Plato 1892)

The Greeks had their epic poetry, but the Egyptians had their recorded histories and their ancient science. Greece had fanciful myths, but the Egyptians had detailed eye-witness accounts of events and an understanding of their scientific causes.

The aged priest tells Solon that this story about "Phaethon" is about a group of cosmic bodies that journey through the heavens over long periods of time on great cyclical orbits and predictably come near enough that some enter the atmosphere of the earth and burn up much of our planet. Moreover, it happens on a known cycle.

Such long-period cosmic bodies that set fire to earth would seem to be comets, even though they are supposedly just cold, icy bodies. But, of course, the 1908 impact in Siberia proved that a comet's frozen volatile gases can explode like an atomic bomb when it enters the atmosphere at extreme velocity. Comets are the fastest objects in the solar system, and therefore, they generate far more explosive energy upon colliding with the earth (or its atmosphere) than do other, slower-moving objects.

In view of Egyptian disdain for Greek science, it is doubtful later Greeks grasped what the wise old priest was saying. The story came to us via these ignorant Greeks: Solon, his interpreter, Solon's friend (Critias' great grandfather), and Critias, and finally Plato himself. Even 19th century translations of Plato's account were not informed by our modern understanding of space science. Only now can we see what the priest meant!

The Egyptians claimed to know of many such conflagrations caused by the same group of orbiting objects. The priest was talking about a specific cluster of objects, or comets, that came perilously close to the earth on a regular, long-term, orbit. It seems that on most returns the cluster had one or more comets strike the earth or explode in our atmosphere. He says this cluster is so predictable that the Egyptians had managed to determine "the long interval" between its fiery returns:

> ...and then, at the usual period (of time), the stream
> of pestilence from heaven descends (down upon the
> earth), and leaves only those of you who are illiterate
> and uneducated, and thus you... know nothing of
> what happened in ancient times... (Plato 1892)

Comets are indeed predictable. They also tend to leave behind a trail of very tiny particles that get strewn around their orbits so that the earth passes through this trail and generates meteor showers on specific days every year. In other words, the same cluster of comets would come by on the same days when it returns. But in the meantime, it leaves behind a dust trail that produces annual meteor showers on the same dates every year. So, the predictability comes down to within a few days.

Now we can see why the time of year and the locations of the comet impacts are so similar in the 65-million year ago extinction, the 15,000-year ago event, and the June 1908 comet explosion in Siberia. All of these comet impacts could be coming from the same swarm of objects in the same orbit that intersects the earth's orbit in late June.

Unfortunately, if that is the case, we are not dealing with any ordinary comet or cluster of comets. This swarm has apparently maintained its integrity for millions of years, in spite of what must have been substantial long-term erosion of its debris field. To explain this, the initial amount of material must have been huge, on the scale of a planet, if we estimate a starting total for all of its comets and associated dust and debris.

The gravitational attraction required to keep this weighty gathering together for such a long time must be colossal. A substantial planet-sized mass is needed, lurking deep in the heart of the swarm, to hold all that material captive in this comet cluster.

To put this into perspective, if all the tens of thousands of asteroids currently in the asteroid belt were gathered together into one object, it would not be enough to form even a small planet like Mercury. On the other hand, most of the material that was once in the belt, c. 99% of it, has long since been ejected (primarily into the outer Kuiper Belt), and only a relative few asteroids are left. (Trujillo 2003)

How could the Egyptian priest of Sais have known the true orbital period of a marauding band of comets that only returned after "long intervals" of time?

It seems the priest was not exaggerating when he claimed the Egyptians had kept detailed astronomical records for thousands of years. They must have directly witnessed at least three visits of the swarm. Otherwise, how would they know it was a repeating cycle? Three visits are a minimum. Four would be needed to verify the cycle.

So, how old was Egypt? It is known that Egypt's first agricultural grain was cultivated *c.* 18,000 years ago, on the west bank of the Nile, near the Faiyum. We also know this came in reaction to a great famine that began 20,000 years ago.

This same time-frame matches up with the period ascribed to the pre-flood "gods" of Egypt. We must remember, however, that the ancient Egyptian chronology known to the priest of Sais is NOT the same as that attributed to Manetho, who wrote three centuries later, after Egypt had been conquered several times, lastly by the comparatively ignorant Greeks. And even Manetho's work, as it has come down to us, is but a shadow of Manetho's actual text, and has been reconfigured by writers of the early Christian era.

What we can determine from our surviving fragments of Manetho is that Osiris ruled in the middle part of the epoch before the flood, the one which began with the rule of Geb, the Egyptian version of Adam. Thus, Osiris ruled Egypt halfway through the old Adamic age that began, by the seven-fold Genesis chronology, about 21,141 B.C. and ended about 9549 B.C. The mid-point of that age was *c.* 15,345 B.C., roughly 17,350 years ago.

Osiris was said to have come into the Nile River valley with his family during a severe famine that had reduced Egypt to cannibalism that we must date before 17,350 years ago. The great famine that began 20,000 years ago in Africa is known to have eased after someone introduced the cultivation of grain *c.* 18,000 years ago. And we know that Osiris was credited by all the Egyptian traditions with introducing farming to Egypt on the west side of the Nile River, the same time and place where scientists found the earliest such grains. This depends, however, upon viewing Manetho's surviving records in light of the sevenfold chronology of Genesis. The seven-fold Genesis record easily restores all these details of history to their real time-frames and can be independently verified by archaeology.

One of the requirements for reconstructing this lost history, of course, is that we identify Osiris (Asar in the Egyptian language) with his biblical counterpart Cain. There is additional evidence for this identification from the *Book of Jasher*, in which we earlier found references to pre-flood famines and the destruction of a third of the earth during the reign of Enosh, the grandson of Adam and son of Seth. The *Book of Jasher* recounts how Cain first went up out of Eden at the time of his banishment. The Hebrew of the *Book of Jasher* text could have been translated as follows (in my alternate English rendering):

> And went out Cain at the time from (the) face of YHWH, out from the place. Asar was (his name) or "out from the place (where his name was "Asar"), and he (Cain/Asar) went moving (about) and wandering in (the) land east of Eden, he and all Asar (had) with him. (*The Book of Jasher* 1840, 1887, 1990, Translator's Preface, p. v)

Just as with the book of Genesis itself, one could have translated this passage differently: "And Cain went out at that time from the face of YHWH, out from the place that he was (but "Shem" (Hebrew name) must be left un-translated), and he went moving and wandering in the land east of Eden, he and all that (were) with him." (*The Book of Jasher* 1840, 1887, 1990, Translator's Preface, p. v) To read it this way, we must omit the reference to Cain's "name" ("Shem") when "Asar" is first used in the text.

Once again, the story of Cain as Asar (Osiris) was being hinted at in a way that could always be explained away if the text fell into the hands of the Egyptians. But why should they continue to conceal this secret for centuries thereafter, even when they were not in Egypt anymore? First, they feared to offend this powerful nation on their western border. Second, they worried that allowing the Jewish people to identify Cain as Asar, that is, the "Egyptian god" Osiris, could lead to a dangerous fascination with paganism.

So, the scribes continued to hide these alternate readings of the text. But as time passed, bits and pieces of the original body of secrets became lost. Most of this knowledge had been handed down orally by memorization and was never committed to writing. Much was lost when rabbis who knew portions of the secret tradition died without being able to pass it on to trusted pupils.

This has been explicitly lamented in regard to the *Kabbalah*, the oral tradition that embraced the tithed in the first half of Genesis and Ezekiel (the tenth book, hence also tithable). (Jeremias 1975, p. 236-238) The "dangerous" references to Osiris are found in chapters four and six of Genesis and in chapter eight of Ezekiel, where the pagan worship in the Temple is described, in detail! Yet, the rabbis told outsiders that their secrets were really about the creation and the Chariot (the throne of God in Ezekiel). (Jeremias 1975, p. 236-238) However, the Chariot is also in the Eighth to Eleventh chapters of Ezekiel. Hiding one topic required hiding the other.

I cannot emphasize enough that Hebrew, by its nature, is a multi-layered language. The reader interacts with the text. We are allowed, actually required, to read multiple levels of meaning in a passage. The English language habit of restricting the text to just one layer of meaning literally blinds us to the full content originally intended to be gleaned from a Hebrew passage. It is like telling a joke based upon a pun or play on words, and no one laughs, because no one considers anything but the "literal" meaning. They fail to "get" the joke. In the same way, some supposed "literalists" directly contradict the Bible's own declarations that it has several layers of meaning (Isaiah 28:9-13, Revelation 13:18, Matthew 13:10-17, 34-35, II Peter 3:8, 16, Romans 11:7-11, 33-36, II Corinthians 12:1-4, *etc.*)

The story of human origins and of the cycles of the ages was such a closely-held secret that the rabbis kept it secure for thousands of years. (Jeremias 1975, p. 239) Only when the rest of the world began to discover their secrets did some rabbis relent, publishing what they knew of these hidden meanings. Today, there may be little left of the lost wisdom of the rabbis that has not been largely revealed or independently discovered.

In any case, we can now use Genesis as a way to organize historical information over the past 23,000 or so years, that is, back to the time of Adam. Genesis gives us, when we realize the great expanse of time it really covers, a framework within which we can understand information gathered from science, mythology, and parallel materials like the works of Josephus and the non-canonical books of *Enoch, Jasher*, and *Jubilees*. The mysterious lost realms of history now become comprehensible.

Nothing better illustrates this new understanding than Egypt. For the first time, we can see that the nation of Egypt had existed 18,000 years ago, that it had been peopled before Osiris and his family arrived to rescue it from a terrible famine, and that Osiris was an exile born long before that, as Cain, in the family of Adam, 23,000 years ago.

The layers of the mystery are now made bare for anyone to see. Osiris/Cain kills his brother, is exiled, and a great famine occurs. A great drought, especially in the Arabian desert and Africa, began quite suddenly 20,000 years ago.

But for those seeking even more precision, we found that the Mercury close-approach Sar cycle placed a Mercury-caused famine right at the front door of Adam, passing near the earth *c.* 20,349 B.C., near the time Cain would have murdered Abel.

One cautionary point about Mercury's proposed 3,600-year Sar cycle: That 3,600 is a round number. Either Mercury (over a long time) was "gravitationally locked" into the earth's 360-day orbit, or "3,600" years is only an approximation! Therefore, either the 3,600-year cycle only applied when earth had 360-day years or it is not a precise quantity. If it were based upon a 360-day year, then it would now be about 3,548.33 of our current 365.24-day years. That falls almost 52 years short per cycle. Curiously, 52 years was a unit of measure in the so-called "Mayan calendar" cycle.

In either case, then, the 3,600-year Sar cycle is not a valid current cycle. Zecharia Sitchin's theory that Nibiru is still on such a 3,600-year Sar cycle, accordingly, cannot be correct. At the very least, the cycle would be 52 years shorter, as we count time today.

Alternatively, if Mr. Sitchin had been right, his proposed planet Nibiru would have come by again, during the current Christian era, but many centuries ago! That obviously did not happen. Sitchin never really explained that fundamental flaw in his theory.

But if the term Nibiru were used only in reference to Jupiter or its "messenger" Mercury, as Semitic language scholar Dr. Michael Heiser has shown, then we already have a built-in explanation for the "end" of Jupiter's 3,600-year cycle: It depended upon the earth having a 360-day year. Once the earth's orbit changed to a 365.24-day year, the Jupiter "Sar Calculation" no longer worked out to 3,600 of our longer years.

On the other hand, Mercury's old orbit is clearly changed from the earth-crossing path it once took through the heavens, when it had literally collided with earth and Mars. The two collisions with the earth occurred very early in the history of the solar system before the 3,600-year Sar cycle began. But its collision with Mars was recent (c. 700 B.C.).

The question is, if Mercury and Mars had been in some sort of stable relationship with the earth for a long period of time, why would all three suddenly get re-shuffled into the new orbits in which we now find them?

These three planets had been pushed around by something else, by another object big enough to reconfigure the inner solar system. Could this interloper have been the same large planet-sized mass that the Egyptian priest of Sais had warned Solon about, the one that is carrying a cloud of comets on a long cyclical journey though the heavens?

We know Egypt has been around for roughly 20,000 years, and its calculation of the orbit of the comet cluster required a minimum of three or four encounters with it. So, the longest time the cluster's orbit could have, if already determined by the Egyptians in 590 B.C. (when Solon was at Sais), was about 5,000-to-7,000 years. This is a fairly broad range, but it does encompass the "Mayan" 5,200-year 360-day cycle.

Could the Mayan cycle have been transmitted to them from the Egyptians?

First of all, we know that the Mayans got this calendar from the preceding Olmec culture, whose origins are somewhat obscure, but went back at least to the time of the Exodus and may have reached back to "at least 3000 B.C." (Gilbert and Cotterell 1996, p. 160) That could mean that Olmec culture began in Meso-America at the same time that Menes was taking power over Egypt and the eastern Mediterranean c. 3100 B.C.

We noted earlier that there had been a widespread belief in a cataclysm or flood around 3100 B.C., apparently not long before Menes went out conquering his part of the world. How much of the world was ultimately subdued by Menes, who was a black ruler from well up the Nile in Africa? Could he or his fellow southern Egyptians have reached as far as the Americas? Were there black explorers in the Americas during Olmec times?

Yes. Exploring near Olmec cities, archaeologists unearthed giant stone sculptures of the heads of what are clearly black Africans. These are not the faces of sailors who got lost. They are almost god-like, the faces of dominant warriors or conquerors. These are faces that could easily be those of Menes' men, or even of the ruler himself. In fact, Graham Hancock has suggested that the face of the Sphinx, before it was later re-carved, may have originally been a black king resembling one of the Olmec's giant African heads. (Hancock 1995, pp. 118-131) Hancock, who had gone to the Olmec region and had personally inspected the heads, argued that they each depicted specific individuals of black African origin. (Hancock 1995, p. 131)

So, specific African conquerors were honored with gigantic stone head sculptures by the Olmecs. There can be no doubt that at some point a powerful African culture with sea-going vessels had dominated Meso-America. From where in Africa had they come?

Menes had sailed north out of the African Nile valley, subjugating all of Egypt. Then, it is thought, he sailed the eastern Mediterranean and subjected all the islands and sea-coasts, including Crete, heart of the Minoan empire, founded at that same time by an Egyptian king named Minos. As noted, Minos may have been the same person as Menes of Africa. The "-os" and "-es" were local grammatical endings; the two names were otherwise identical.

There are other links to Menes, specifically involving standing stones. Off the northwest coast of Scotland, the Isle of Lewis has a well-known circle of standing stones aligned to the solstices. The tradition of Scotland is that the Isle of Lewis stones were set up by Africans far back in the Bronze Age (possibly *c.* 3000 B.C.). In Brunswick, Maine, across the Atlantic, there is a set of standing stones that seem to have used early Egyptian cubits. People who observed both the Scottish stones and those in Maine contend that they look identical. Still more standing stones of similar design are found in West Virginia. These, along with the stones on the Isle of Lewis off Scotland, and the stones of Maine, form a great circle on the earth, as if erected at a time when the earth was tilted differently than it now is. The standing stones may show signs of being reworked after an axis-tilt change. If one follows the great circle further south, it leads directly into the heart of the Olmec empire, right to where the giant African stone heads were found in the jungle of Mexico.

There are also standing stones in Africa, well up the Nile, in the area from which Menes came. All these standing stones might be artifacts of Menes' empire. If they are, it is interesting that the ancient name used for these "long, tall stones" in Celtic lands was "Menhirs." (Pritchard 1969, p. 820a) The "Men-" root meant "stone." The "Dolmen" or table-stones of France, Britain and New England share the same linguistic root: "Men-." (Pritchard 1969, p. 389a) These "Mens" ("stones") were set up by Africans in the days of Menes.

Is there any more specific evidence linking Menes/Minos and his African naval empire with northwest Europe and the Americas?

Absolutely... British naval historian Gavin Menzies argues that the Minoan empire was in fact a global one, mining metals in northern

Europe and the Americas. He cites metallurgical analyses of iron from Michigan and the Minoan lands, as well as DNA evidence proving that the Minoans had personally visited the Americas. (Menzies 2011)

Finally, cocaine from Meso-America has been found in the tombs of several Egyptian pharaohs. Thus, there can be no doubt that the Egyptians were fully capable of sailing to Olmec lands, that black African warriors dominated the Olmecs, that a Minoan-Egyptian empire was exploiting American resources, and that the Egyptian rulers traded with the Olmecs.

The string of standing stones that the ancients themselves said had been erected by Africans, that they even named "Men-hirs," extending from the birth-land of Menes to France and the British Isles, to Maine and New England, down the coast, "aimed" right at the capital of the Olmec empire on the Gulf of Mexico, could well have been a trade route pioneered by the vessels of Menes himself. Perhaps one of those stone heads is the very face of Menes, the black African ruler who united Egypt and conquered the world, establishing the empire that formed the basis of our modern mercantile civilization.

A key part of lost human history can now be reconstructed: Menes was not just a king; he was an explorer. He inspired his people to sail a fleet north, up the Nile from southern Africa in *c.* 3100 B.C., apparently prepared to explore the coastlands of the known world. He must have constructed a great fleet of sea-worthy vessels and outfitted them with substantial supplies for his great expedition.

But why would he have felt a need to do any of this? Such an undertaking had to be fueled by more than casual curiosity.

We know that the ancients believed there had been a flood at this time. The Egyptians regarded the Nile River as an escape route that allowed them to flee up-river to get to high ground deeper in Africa in the event of a flood. In his comments to Solon, the priest of Sais specifically mentioned the Nile's role in protecting Egyptians:

> ...and from this calamity, the Nile, who is our never-failing savior, saves and delivers us... When, on the

other hand, the gods purge the earth with a deluge of water (from above) ...in this country (Egypt) neither at that time nor at any other time does the water come from above... having always... come from below... (Donnelly 1882, 2013, p. 8)

Notice that the priest of Sais says the water comes "from below" whenever there is a flood, not the sky, yet he attributes it to the repeating cycle of comet-like objects. How do cosmic objects from space cause a flood to engulf Egypt "from below?"

The priest was in the Nile River Delta, virtually at sea-level. "From below" could only refer to an inundation that came from a sudden overflow of the Mediterranean Sea in a great tidal wave sweeping ashore. There are several ways this could happen as a result of cosmic bodies overhead:

First, an undersea quake can trigger a tsunami. This quake could be set off by the gravitational disturbance of a sizable cosmic body passing near the earth. Several great earthquake faults lie due north of Egypt, under the sea.

Second, a volcano could be triggered the same way. Volcanoes can explode with stupendous lateral force, sending a wall of water rolling over Egypt. There are volcanoes in the eastern Mediterranean region, also due north of Egypt (*i.e.* Thera).

Third, a comet or some other large chunk of cosmic debris could strike the Mediterranean Sea and directly cause a tidal wave.

Fourth, an impact in the Atlantic could sweep the ocean into the Mediterranean basin, similar to the events of the Deluge of Noah, almost 6,500 years before Menes.

In all cases, those in the low-lying northern parts of Egypt would have been swept to their death, unless they had warning and had time to flee southward to escape. No doubt such warnings and evacuation plans were part of life in the north. The priests were skilled at making predictions of cosmic events. This explains the priest knowing the true periodicity of the comet swarm and claiming they had time to escape. It also explains Joseph warning his brothers, "God will surely visit you" (Genesis 50:25), a prediction fulfilled when Mercury, "the angel (messenger) of the Lord," came by at the Exodus.

The peoples of the south would not necessarily have welcomed the sudden influx of northern refugees. Menes and his kingdom must have had mixed feelings when many thousands of these homeless northerners poured up the river into their territory. Tales of terror came with the refugees.

Menes had two options: He could house them all permanently or he could outfit a great fleet to carry them back to their northern homeland after the danger was over. His fleet could anchor in the Nile River Delta region, providing the refugees food and housing until they could rebuild their ruined, mud-buried villages.

Now we see why Menes built his fleet and stocked it with months of provisions. To make sure it went as planned, Menes led the expedition and took warriors with him. The north may have recovered much faster than he expected. Menes was left with a fleet of vessels and enough provisions to go out into the Mediterranean seeking other people in need of help. It may also be that some refugees on board were originally from the islands and nearby coastlands. Given the historic role of Egypt as a merchant hub, that is almost certainly the case. Menes proceeded to help these people reconstruct their coastal towns as well. The refugees would have been extremely grateful to him and his people.

Menes was certainly a brilliant leader. He knew that reestablishing all of these peoples in their own lands would lead to military and merchant alliances with the grateful survivors. He was in a position to be everyone's benefactor. He was creating a huge debt of gratitude that all of these devastated peoples could never repay. He would be forever remembered as the man who rescued the world from the cataclysm of 3100 B.C.

Echoes of Menes global mercantile empire may have survived far longer than most people today realize. Not only do the words Menhir and Dolmen preserve his fame, but so do many other modern words.

For example, because the Minoan empire engaged in the copper trade and made copper and bronze coinage popular, we have a possible explanation for the origin of words like "mining" and "mine" and "money." Terms like "minimum" and "minus" and "minor," among many others, could have a similar origin. The words "moan" (due to heavy taxes) and "munificence" (generosity) may also derive from these.

Moreover, over the long 2,000-year history of the Minoan empire (from 3100 B.C. to 1100 B.C.), it is known that their written language changed direction from a right-to-left direction (like Hebrew), to a left-to-right direction (like Greek and English). That mid-course reversal of their writing direction might have led to considerable confusion, producing a lot of reversed words or syllables. In this way, the "Men-" root may have also been transposed into a "Nem-" or "Num-" form.

This could explain the origin of words like "number" and "numerous" and "numismatics" (collecting coins).

Other conceivable derived words might be "man" and "woman" and "mankind" and "human." Or "name" and "nominal" and "nominate" or "denomination."

More proper name connections might have produced "Nemesis" or "Nome" (the enumerated classification of districts on either side of the Nile in Egypt).

While we are discussing such ancient word derivations, there is a strange story that we should mention in regard to a possible forgotten connection between Menes and "Memnon." When tourists visit Thebes, it is often the southernmost part of their trip to Egypt. On the west bank of the Nile across from Thebes sit two 60-foot high monolithic statues called "the Memnon Colossi," which modern Egyptologists attribute to the Pharaoh Amenhotep III. (West 1995, pp. 377-378) Carved out of hard sandstone that had been laboriously hauled from 100 miles further south, the two statues are defaced and in a state of ruin that makes them look almost like alien beings. (West 1995, photos, p. 377)

But in Roman times, the identification of the statues had been quite different. The people of that day thought their inspiration had not been Amenhotep III, but had been "Memnon," an Ethiopian (that is, a black African like Menes) slain by Achilles, the Greek hero of Troy. (West 1995, p. 378) The images must have then looked like those of a black African, although it is not known why he was identified with "Memnon;" they believed him to have been the semi-divine son of Eos, the goddess of dawn. (West 1995, p. 378)

The worn condition of the statues had suggested great antiquity in Roman times. Perhaps they inferred that the person depicted must have been the oldest Egyptian ruler of the current era, namely, Menes, first king of Egypt at the "dawn" of the age that had begun with the flood of 3100 B.C. This led to confusion because of similarity to the "Memnon" slain by Achilles, at least as the names were then represented in their languages.

So, Memnon/Menes was truly the "Manu" or "founder" of a new age of the world?

As we can see, the name of Menes ("Men" in Egyptian) was defined by Mem and Nun. These letters meant "water" and "fish" respectively. (Quaknin 1999, p. 246, 258)

An ancient belief held that the alphabet embodied a hidden astrological design, that each letter represented a constellation of the heavens. The consecutive letters Mem and Nun represent "the Water-Jar Man (Aquarius)" and "the Fish (Pisces)." Originally, Pisces and Aquarius were one great 60-degree sign: the "Fish-Man." The specific "Fish-Man" in question, the one honored with a pair of constellations, was Noah, the "Manu" or surviving "Fish-Man" of the post-flood age (hence, the silly Hindu "talking-fish" story).

It was said by the Jewish historian Josephus that Seth established the priesthood which originally named the stars. We are also told that at about this time, "men began to call upon the name of the Lord," asking for God's help. (cf. Genesis 4:26) There are no atheists in fox-holes. Perhaps a terrifying event had intervened after the birth of Seth's son. It may have frightened the men of that day so much that they repented and begged God for help. Could this event have been an early passage of the great cluster of comets?

If a cosmic calamity had been the reason for calling upon the name of the Lord for protection, then the naming and organizing of the divisions of the heavens was likely undertaken in order to be able to predict and prepare for the next such event.

The "Fish-Man" or "Manu" became a title that meant "Rescuer" of his people. Thus, Noah became a type of Christ. And the Egyptian Menes, a black ruler from southern Egypt, who might otherwise have lived out his life in obscurity, rose to the occasion after a great cataclysm, rescued not only Egypt, but many other flooded coastal regions. For this, he was honored with the title "Fish-Man"/"Rescuer" (Menes).

It may be that the standing stone arrays were set up by him, using a design native to his homeland. Possibly it was a device for providing the locals with a proven Egyptian technology for predicting future cosmic calamities. If so, how did it predict them?

One of the largest and by far the most imposing of those set up at that time, in 3100 B.C., was Stonehenge. (Knight and Lomas 2001, p. 145*ff*) It is unique in that its precise location is where the day of the summer solstice in late June (remember the summer solstice's cosmic impact significance) will be exactly sixteen hours and the night will be exactly eight hours. (Knight and Lomas 2001, p. 146) Moreover, the standing stones at Stonehenge, although constructed of truly massive pillars of granite, are precisely oriented to monitor the summer solstice. (Knight and Lomas 2001, p. 146) This fact alone shows us that in 3,100 B.C., they felt a need to pay strict attention to the very time when we know the comet cluster comes by, in late June.

But there is more to it than that. The giant stones were hauled a great distance so that they could be erected in this particular spot, where the summer solstice is perfectly timed. The effort made by the builders to place the most durable available granite stones in this specific location reveals that they were deadly serious and needed the stone array put exactly here to maintain its summer solstice alignment for thousands of years.

The time of the summer solstice was their primary concern. Something that they had experienced at that time was motivating them. This must mean that there had been a summer solstice cataclysm in *c.* 3100 B.C. And, they needed their cataclysm warning device, the Stonehenge array, to be so sturdy and unmovable that it would keep its orientation for several thousand years so that it could warn a future generation.

Thus, like the Egyptians, the builders of Stonehenge used huge blocks of granite carved out of the ground with great difficulty and moved over substantial distances. They did this in order to leave some future generation, thousands of years later, a way to predict a coming cosmic cataclysm tied to the summer solstice time-frame.

That was 5,000 years ago. Much of the Stonehenge complex is fallen down, as are the stones in most of this widespread culture's other ancient constructions. (Knight and Lomas 2001, pp. 148-149, 296*ff*) How could these surviving stones be used to predict anything at this point?

Ah, but they can. Megalithic stones that remain aligned to fixed stars in the sky or to the sun continue to provide an early warning for one particular kind of cosmic foe, a problem that justified warning mankind thousands of years in advance: Stonehenge, like all the builders' other megalithic devices oriented to fixed heavenly bodies like the sun, can be used to detect gravity waves.

Any disturbance to the gravitational field of our solar system will misalign our planet's relationship to the stars of the heavens and to our own sun. The precise moment of the solstice, the exact angle of the solstice sunlight at dawn or sunset, or any other such measurable event, can reveal that a gravity wave is passing through the earth. The closer the source is, the greater the disturbance. The faster the source is moving, the greater will be the effect. And, of course, the more massive it is, the bigger the influence it will have.

The ancients were worried about all three things. They set up stone monoliths to monitor the timing of the solstice, its location, the rotational rate of the earth along a great circle route, and any other variable they could measure. All this is very revealing.

Whatever it was, it must have been reckoned extremely massive. Moreover, it must be moving at a high velocity. And it is able to approach very close to the earth.

Needing a megalithic detector reveals that otherwise they might not see it soon enough to flee. This implies that it is very dark and does not give off much light, if any.

The attention devoted to the summer solstice reveals that, in spite of the enormous effort they expended, they did not expect to get more than a few days of warning. The events we have been studying came within ten days of the solstice. These few days must have been regarded as a matter of supreme importance to those facing the threat. They did all this just to give some future generation only a week or so to flee a "Dark Intruder."

Of course, the earth itself is the real gravity detector. All these stones were merely a way to detect changes in the position of the earth. Our planet is going to be jarred out of its normal rhythms by what is coming. The stones merely provide a handy way to watch for small disturbances before it is too late. The summer solstice is the best time to make these observations because it is close enough to the arrival of the "Dark Intruder" that its approach will have by then become sufficiently detectable.

The "precision" of the Mayan calendar, as we currently interpret it, depends on how well the Mayans themselves understood it. If they or the Olmecs had miscalculated it, then our estimated dates based upon their inscribed dates will also be in error.

In other words, even if the overall cycle length were correct, they may have gone off track at some point in recording dates down through the centuries. The very fact that some think the calendar was supposed to expire around December 22, 2012, shows that it has been misinterpreted by someone. All of the other evidence we have seen is firmly centered upon the period around the summer solstice, not the winter solstice. The true ending time of the calendar is summer, as even some Mayan scholars have argued.

But the value of the calendar is to usher in the time-frame when one should begin to focus on the megalithic detectors. From that moment on, it is no longer a countdown. It is all about measuring carefully to determine if there is the slightest change in the expected solstice parameters. If anything is not as it should be, the end, we must assume, will be only days away. We may not be able to see it yet, but it will be out there, coming at us in the dark, moving faster and faster, closer and closer, unfathomable, unstoppable.

So, we have growing evidence for a visit from this "Dark Intruder" near 3100 B.C. Is there more proof that something descended from the sky to devastate the earth at that specific time?

Yes, indeed Professor Ioannis Liritzis of the University of Rhodes (Turkey) found many lines of evidence showing that a sizable comet had impacted the Mediterranean Sea in *c.* 3150 B.C., as reported by Christopher Knight and Robert Lomas in *Uriel's Machine,* their valuable investigation of megaliths and comet impacts. (Knight and Lomas 2001, p. 296) And Knight and Lomas themselves independently determined that such an event had occurred around c. 3,150 B.C., based upon magnetic anomalies and nitric acid peaks in ice-core samples that serve as signatures for a comet impact. (Knight and Lomas 2001, pp. 58-64, 296)

There is more. In 2008, two British scientists: Alan Bond and Mark Hempsell, who were researching European impact craters dated around 3123 B.C., found an eyewitness account of a large cosmic body that passed over the Euphrates basin at that time. It was recorded in a Sumerian text inscribed on an Assyrian tablet from Nineveh. The tablet, known as the "Planisphere," is circular, partly illegible, but rather notorious among archaeologists, who consider it to be an extremely mysterious astronomical chart of unknown meaning. (Haines 2008) (Sitchin 1985, pp. 272-282)

Zechariah Sitchin attempted his own novel translation of the puzzling "Planisphere tablet," contending that it described "alien gods" from the planet Nibiru landing a "fiery rocket" by flying northwest up the Mesopotamian Valley. (Sitchin 1985, pp. 272-282) The British scientists disagreed, however, arguing that this object was a comet burning up as it fragmented in the atmosphere.

Note the north-westerly motion of the object. This was the same north-westerly impact pattern we saw in the 15,000-year ago comet that broke up over North America, scattering its fragments along a path from Chicago to Siberia.

In their 2008 investigation of the event, the British scientists said the Sumerian eyewitness had watched the object moving through the pre-dawn sky, passing from southeast to the northwest, before going out of sight over the Turkish mountains. It was said to have been a large "bowl-shaped" object enveloped by smoke.

The "bowl" shape of the object is extremely odd. Its size had to be huge for the witness to have been able to discern a bowl-like contour hidden inside an enveloping cloud of smoky debris just before dawn. Anyone who has seen a large meteor or bolide burning up in the atmosphere will realize how difficult it is to determine the shape of the object beyond an assumed spherical form, given the few seconds it is visible (at most) before it flames out and disappears. And such objects are not enshrouded in smoky debris clouds. They are brilliantly illuminated as they burn up, not hidden in smoke.

Unlike an object burning up as it enters the atmosphere, this object seems not to have been in flames. Rather, it was concealed inside a dark cloud of smoke-like debris that apparently had a gap in it large enough to view the massive object within. That is hardly conceivable if it were burning up from air friction. Any debris cloud of smoke would be trailing behind it in a tail. A superheated fireball engulfs an object entering the atmosphere, masking the shape it has. The eyewitnesses of the 1908 Siberian object's last moments described seeing "an enormous fireball" and a series of explosions as it came down early in the morning of June 30th. (Knight and Lomas 2001, pp. 48-49)

But, that is apparently not what the Sumerian account described happening in the comet event from 5,000 years ago. The object seems to have moved along slowly enough that a bowl-like shape could be detected, not hidden by a fireball, but visible inside a cloud of smoke that hovered around it.

In other words, there is nothing in this description that is consistent with an object moving within the atmosphere. But the Sumerian description makes perfect sense if the object were outside the atmosphere, not burning up, but simply accompanied by a smoky cloud of debris that was gravitationally bound to it.

If it were outside our atmosphere, then the "Planisphere" description requires that it be a far larger object, not an ordinary comet, moving slowly, but steadily, past the earth. The observer has time to peer through a gap in the debris cloud and see a shape he describes as a "bowl" or a hollowed-out, concave, but spherical object.

The hollowing out or crater-like gouging of a spherical cosmic body indicates that it has been damaged; that is, it had been in a collision with another large body.

A plunging comet streaking overhead en route to a fiery demise will not display a massive crater wound of this kind. Such an impact would have pulverized a comet. It should not leave anything behind, much less a giant gouged-out bowl of a crater. Such a crater requires a hard, dense object, not a comet formed of dust and ice and frozen gas.

Therefore, the Sumerian witness did not see a comet burning up before it plunged into the Mediterranean Sea. Rather, he was watching a massive, dense cosmic body moving much more slowly by the earth, slow enough to allow the witness to observe its peculiar shape through its surrounding cloud of dark debris. He may have seen some "embers" of what seemed to be fiery "sparks" darting out from it as it moved by, but these could have been actual comet-like debris orbiting it and becoming illuminated as it passed around the dark side of the object and emerged into the sunlit side.

The impact that gouged-out the bowl shape explains why the massive body was accompanied by this great cloud of comets and other kinds of collision debris.

The two British scientists were able to glean one more revealing detail from the tablet: the event had taken place in the pre-dawn hours of June 29, 3123 B.C. This precise a date provoked criticism from some Sumerian scholars. No one was supposed to be able to be that precise. Yet, a date in late June is exactly what we have come to expect. (Haines 2008)

On the other hand, it may be that this year 3123 B.C. (nine years before the Mayan "calendar" starting date) is subject to revision. It would take something like 627 years of 360 days following this event to explain these 9 years' difference. Curiously, those 627 years subtracted from 3123 B.C. would bring us back to 2496 B.C. In other words, it covers almost exactly the difference between the Septuagint's revised date for the Deluge and the Hebrew text's date for the Deluge. It may be just a coincidence, but if not, then it might suggest that something had changed the old 360-day orbit of the earth around 2496 B.C.

This could be Mercury, if it had come by in 2496 B.C. (147 years earlier than its expected 2349 B.C. fly-by on its 3,600-year cycle of close approaches to the earth). The passage of the "Dark Intruder" in c. 3123 B.C. may have interfered with Mercury's orbit.

The Dark Intruder hurled several comets at the earth as it flew past 5,000 years ago, c. 3123 B.C. One hit the Mediterranean, setting off a towering tidal wave that swept over the Nile Delta and surged well upriver. Refugees had been warned in advance to evacuate the low-lying area of Lower Egypt in the north. They raced into the highlands of Upper Egypt where Menes was ruling. After the disaster was over, he launched his expedition to restore the north, and then sailed out into the sea to resettle the Mediterranean basin, which the tidal wave had mostly scoured clean. His fleet eventually followed the coastline to Turkey Greece, Italy, Spain, and out into the Atlantic to Gaul and the British Isles, then on to Iceland, Greenland and the Americas. At some point, his ships returned to the same areas on a second voyage of inspection and set up detector stones. Also, starting around 3100 B.C., the whole world's agriculture output boomed.

We have now documented three "Dark Intruder" Mayan cycle events: One around 20,000 years ago (the sudden creation of the Arabian Desert). One about 15,000 years ago (the fiery destruction of the American *Nephilim* that opened up the ice corridor through the great ice sheet). And the most recent one, some 5,000 years ago (the 3100 B.C. impacts). In between these last two, if they were part of a 5,000-year cycle, there should have been one other similar event, approximately 10,000 years ago.

Did anything unusual happen at that time?

Yes, 10,000 years ago was a brief time of extremely rapid climate change. The earth abruptly warmed. This came at the end of a 1,300-year mini-ice age that had begun with the sinking of Atlantis, at the time of the Deluge. Also, agriculture went through an explosive increase at that time, 10,000 years ago, that is, around 8,000 B.C.

According to the Mayan calendar cycle, the date of the comet event ought to have been c. 8,250 B.C. However, radiocarbon dates that far back have a sizable error rate. Thus, a 10,000 year ago radiocarbon date can actually be off by several hundred years.

In their book, *Uriel's Machine*, Messrs. Lomas and Knight built a case for seven comets suddenly striking the earth at a time they identify with 7640 B.C. (Knight and Lomas 2001, p. 58) However, their case for this cluster of seven impacts includes several pieces of scientific evidence which they admit might have actually been dated to c. 8000 B.C., or even somewhat earlier. (Knight and Lomas 2001, pp. 55-68)

The evidence they uncovered included rains of tektites (glassy meteoroids), burn layers, radiocarbon anomalies, nitric acid peaks in ice cores, magnetic field disturbances, and so forth. (Knight and Lomas 2001, pp. 55-68) There can be little doubt that something terrible occurred about that time, 10,000 years ago.

The priest of Sais mentioned that his city was then about 8,000 years old, which meant it had been founded around 8600 B.C. This was an approximate age, however. So, it may be that the city had actually been founded around 8250 B.C. If there had been a time of wide-spread devastation, new cities would no doubt have been established afterward, just as we saw during the global revival of civilization in c. 3100 B.C.

Were any other cities or towns either destroyed or founded at that time?

It happens that we now have several new archaeological discoveries covering this specific period. They detail the sudden appearance and the abrupt end of a number of cultural settlements at that time.

For example, on the highlands of Turkey near Edessa (now known as Urfa) is an ancient stone religious site called Gobekli Tepe. It is dated from 9600 to 8200 B.C. (Mann 2011) That places right it in the 1,300-year gap between the Deluge and the event of 8250 B.C.

Gobekli Tepe is on the natural "downhill" route from the mountains of Ararat in Turkey, where Noah's ark is said to have rested. Thus, this is exactly the area where Noah's descendants should have settled not long after the flood. The valley below would have taken much longer to dry out than the rocky hill country of Gobekli Tepe. So they would have to have halted their migration at about this point.

The dig at the site found that it had been deliberately buried, around 8200 B.C., by the people who used it. (Mann 2011) If they had foreknowledge of a cataclysm that was about to occur, this burial would make perfect sense. Given the apparent ancient function of stone megaliths as devices for detecting the Dark Intruder, it is significant that Gobekli Tepe was built of megalithic standing stones arranged in circles. Indeed, it is difficult to find any other purpose for these stone arrays at Gobekli Tepe.

Of course, the "official" version is that it was the birthplace of religious ritualism. Gobekli Tepe had symbolism, but it is only religious if one defines "religious" in a very broad sense. There were carvings of animals on the sides of some stones, and the monoliths themselves seemed to suggest human figures. But today one can find pictures of people and animals at zoos and museums that are in no way religious. Of course, a stone circle with zodiacal (animal) symbols could be astronomical, rather than religious.

Trying to make sense out of Gobekli Tepe, the German excavator decided that the builders were continually burying the older generation's stones and then erecting new ones of their own, but with noticeably less skill each time. This view of the events does not make much sense, even to the excavator. (Mann 2011)

On the other hand, if the stones were gravity detectors with alignments to the sun, moon, and stars, the builders would have gradually realized there was no real need for artistic carvings or other 'design' elements. The rather crude-looking standing stones of Callanish, on a desolate island off Scotland, for instance, measure the solstice quite well, without any need for nice, neat artistic embellishment.

All around Turkey and south of it are the remains of twenty ancient settlements dating between 13,000 B.C. and 6000 B.C. That is, they begin right after the great comet bombardment of 15,000 years ago and they end at the time of the Mercury passage of 6000 B.C. Right in the midst of these two clear demarcations in the dating sequence are some additional distinct breaks in the archaeological record.

The most dramatic is the one 9600 B.C., when Atlantis sank. Nine of the ten settlements dated to that period either abruptly end or begin about that time. The other one may simply have been resettled after the flood. The Atlantis legend explicitly identifies nations at war in the eastern Mediterranean just before the 9,600 B.C. "greatest deluge of all," as the priest called it. At this same time, Noah was carried up to Ararat by a tidal wave from the North American tectonic de-leveling event, which swept across the Atlantic and poured through the funnel of the Mediterranean basin, generating a mountain-high wall of water that scoured clean the Turkish plateau. What was left was a lifeless canvas of muddy mountains upon which the survivors would erect post-flood settlements like Gobekli Tepe.

We have enough data to roughly chart human migrations during this time. Before the 9000 B.C. flood, the settlements begin in the Holy Land, exactly when and where we found Enosh had been ruling when the comets hit just before 13,000 B.C. The locations of the towns slowly move uphill and north toward the plateau of Turkey until 9600 B.C. In the period after 9600 B.C., the settlements reverse this migration. Starting near Ararat, the sites begin to increase from family-sized units to real towns and move steadily away from Ararat to the west and south, slowly progressing downhill toward the mud-plagued lowland areas. The progression is clear-cut and unmistakable.

If we were talking about stars, the scientists would cite it as evidence for a cosmic explosion, and they would be pointing at the "constellation" of Ararat as the source of the explosion. It would be the primary focus of their attention. But when dealing with the biblical Noah and the flood, such an investigation is forbidden.

A distinct break in the record occurs 8200 B.C.; some settlements survived or were rebuilt. About 7500 B.C. came another break. Mars has ⅑th the earth's mass; it could have been moved near the earth by the "Dark Intruder." For a time, things were stable, until Mercury's passage $c.$ 6000 B.C. Then everything stops, including other settlements, lost under what is now the Black Sea north of Turkey. Following 6000 B.C., human settlement slowly migrated back down toward what would become Sumer and Babylon well after 4000 B.C. (See chart at the end of this report for more details.)

The Egyptian priest of Sais had told Solon that the object that had burned up the world was the one that the Greeks had named "Phaethon." He said that Phaethon would come by "at the usual period," indicating that he was taking a very casual attitude about its cycle. It appears that, by the priest's time in $c.$ 600 B.C., the Egyptians were confident that they had pinned down the orbit of Phaethon and were not especially concerned about an imminent return at that time. Therefore, by 600 B.C., they must have already witnessed and documented the timing of at least four encounters, which would be the four most recent ones ($c.$ 20,000; 15,000; 10,000; 5,000 years ago). So, the Egyptians had been keeping a record of Phaethon's returns for 17,400 years, as of 600 B.C.

The priest told Solon that the Greeks had lost track of the continuity of history. Yet the priest admitted that the Greeks did recall a garbled version of the event of 15,000 years ago. That means the Greeks should also have had some sort of garbled memory of at least the two more recent events, c. 10,000 and 5,000 years ago.

The Greeks did indeed recall the object's other returns. They called it "Phaethon" 15,000 years ago, but in 3,100 B.C. it was "Typhon." The previous event, 10,000 years ago, was named for Typhon's "father," Typhoeus. The earliest, 20,000-years ago, was called Python. Python and Phaethon were pre-flood ice age events, while Typhoeus and Typhon were post-flood. Note the reversal of the "P" and "T" sounds. This reversal occurred around the time of the Deluge in 9,600 B.C., when the ice age ended.

The return of "Typhon" (its most recent name) was left to the Mayans to predict. Some argued the "calendar" might end at the Summer Solstice of 2011, but their voices were drowned out by "mystics" and "shamans" insisting on December 2012. Mayan stone inscriptions said it would usher in a rain of objects out of heaven, but these would be only the beginning of a time of massive earth upheavals. The priest of Sais indicated Typhon will shower the earth with clusters of fireballs that will set fires on the earth.

Peter Davenport of the National UFO Reporting Center, in mid-June of 2011, was abruptly flooded by global reports of clusters of fireballs. These fireballs have continued in 2012 and have increased in number since that time. We have seen random fires erupt in Colorado, New Mexico, Kansas, and numerous other places. There were even reports of fireball clusters grounding fire-fighting aircraft on the night of the Summer Solstice in 2012.

The leading edge of Typhon's fiery debris field had arrived, right on schedule. Could the Dark Intruder itself be far behind?

Chapter Bibliography

Donnelly, Ignatius. 1882, 2013. *Atlantis, The Antediluvian World.* Seattle, WA: CreateSpace Independent Publishing Platform.

Frazer, Sir James E. 1964. *The New Golden Bough.* Edited by Theodore H. Gaster. New York: New American Library.

Gilbert, Adrian, and Maurice Cotterell. 1996. *The Mayan Prophecies : Unlocking the Secrets of a Lost Civilization.* Rockport, Massachusetts: Element Books.

Haines, Lester. 2008. The Register. March 31. Accessed May 6, 2018. https://www.theregister.co.uk/2008/03/31/kofels_asteroid/.

Hancock, Graham. 1995. *Fingerprints of the Gods.* New York, New York: Crown Publishing.

Jeremias, Joachim. 1975. *Jerusalem in the Time of Jesus.* Philadelphia: Fortress Press.

Knight, Christopher, and Robert Lomas. 2001. *Uriel's Machine: Uncovering the Secrets of Stonehenge, Noah's Flood and the Dawn of Civilization.* Gloucester, Massachuettts: Fair Winds Press.

Mann, Charles C. 2011. "The Birth of Religion." *National Geographic,* June: 34-59.

Menzies, Gavin. 2011. *The Lost Empire of Atlantis: History's Greatest Mystery Revealed.* New York, New York: William Morrow, HarperCollins Publishing Group.

Perry, J. H., trans. 1840, 1887, 1990. *The Book of Jasher.* Thousand Oaks, California: Artisan Sales.

Plato. 1892. Timaeus Dialogue. Vol. III, in *The Dialogues of Plato,* by Plato, translated by B. Jowett. London, England: Oxford University Press. http://oll.libertyfund.org/titles/plato-dialogues-vol-3-republic-timaeus-critias.

Pritchard, James B. 1969. *Ancient Near East Texts.* Third. Princeton, New Jersey: Princeton University Press.

Quaknin, Marc-Alain. 1999. *Mysteries of the Alphabet.* New York: Abbeville Press.

Sitchin, Zecharia. 1985. *The Twelfth Planet.* New York: Avon Books.

Trujillo, Chadwick A. 2003. "Discovering the Edge of the Solar System: Recent Discoveries Suggest that Planets Larger than Pluto may exist in the outer reaches of our solar system." *American Scientist*, September-October: 424-431. http://www.jstor.org/stable/27858273.

Turner, Patricia, and Charles Russell Coulter. 2001. *Dictionary of Ancient Deities*. 1st. New York, New York: Oxford University Press.

Velikovsky, Immanuel. 1950, 1967. *Worlds in Collision*. New York: Dell Books.

West, John Anthony. 1995. *The Traveler's Key to Ancient Egypt*. Wheaton, Illinois: Quest Books.

Chapter 7

The Rise of the "Dark Intruder"

Typhon is a Dark Intruder advancing toward our doorstep, a spectre lurking beyond the night, out of view, sliding quietly toward us, slipping by our hedge with ever-increasing pace. It cannot be stopped, nor diverted. Its approach is silent and inexorable.

The cosmic ghost that haunted the Mayans, now haunts us. It has begun pelting our planet with fireballs. Typhon's debris field is enormous, as this chart shows:

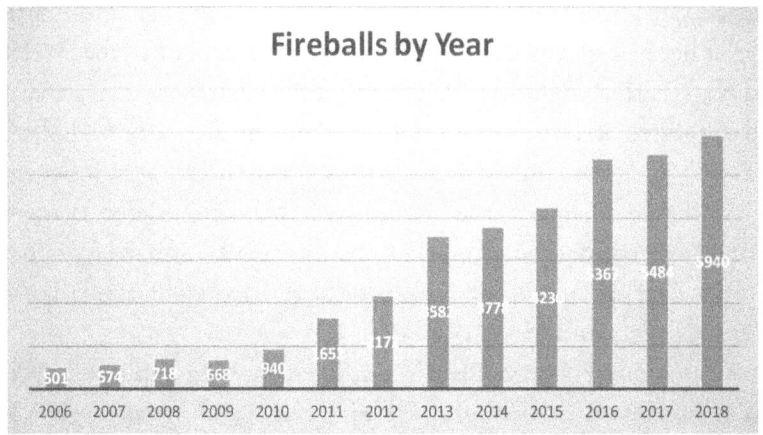

It is easy to see that the Mayans were not anticipating a mere meteor shower when they spoke of "gods descending from heaven on silken ropes" at the end of their calendar cycle of 5,200 years. They knew this was not to be anything like the timid annual meteor displays that scatter a few brief white streaks around the sky over a couple of nights, and then fade away for a year. Those meteors are but grains of sand and gravel burning up in the stratosphere, almost never reaching the ground.

No, the fireballs of Typhon are not so quickly incinerated. They are rocks, some as big as houses, blazing into brilliant fiery glory as they penetrate far more deeply into the atmosphere and threaten mountain forests and homes with sudden immolation.

The air at sea level, as we know, is heavier and denser than in the mountains. Any chunk of debris falling toward the ground has an ever-escalating level of air pressure to plow though. The air slows its descent, which reduces combustion to the point that the object tends to flame out. Accordingly, the risk of actually being hit by a burning fireball decreases rapidly at lower altitudes but increases as one goes up in elevation.

In June 2012, blazes erupted randomly in the high forests of the Rocky Mountains surrounding Colorado Springs, Colorado. Hundreds of homes were taken by sudden fire. Then, on the summer solstice, the authorities were compelled to ground their fire-fighting aircraft because of fireballs raining down over Colorado, New Mexico, and Kansas that night. At first, the random distribution of new fires had suggested arsonists at work. But, when the deluge of fireballs fell randomly over the same sites as the fires in the three states, it began to look as if at least some of the blazes may have been initiated by objects falling from the heavens. Yet, that possibility was omitted from news reports. Fearing panic, the media withdrew from the fireball story. (Cotter, Fellers and McMillin 2012)

The reporters probably did not know how to explain it. What was it out there in the solar system that was hurling a growing number of fireballs at the earth? How could there be so many? Over 29,000 fireballs have been reported over North America alone since 2006, and nearly two out of three have fallen in the last half of that period. The 2017 total of 5,456 fireballs was over ten times the 501-fireball total in 2006. In the past seven years, the annual count is estimated to have grown five-fold. At that rate of increase, one could theoretically project future fireball activity approaching 50,000 a year within a decade of the record set in 2017.

The usual source of meteor showers, a disintegrated comet, could never generate that many bolides and fireballs over such a time-frame. The typical annual meteor shower might have only one or two fireballs, if any. These new fireballs are sometimes being seen in clusters, even numbering dozens of flaming objects at a time.

By the sheer volume of material involved, something very, very big is generating all these objects. Moreover, as we have already noted, the gravitational power to hold all this debris together in a discrete part of the orbit requires a massive cosmic body. But there is a new observation about its size that we can now make.

This debris field has thus far taken seven years (as of 2012) to engulf our planet, as this truly enormous captive ocean of rubble has steadily moved into our solar system.

Imagine, if you will, being in a tiny boat at sea on a completely dark night. That little boat is the earth, which is barely a pebble orbiting the vast blackness of space. But then suddenly, you feel a breeze, not of wind, but the motion of a ghostly ocean liner, abandoned and lifeless, drifting toward you. You cannot see it, but you feel it coming. It is huge by comparison to your little boat, and it drifts along silently, gigantic and massive in the waters beside you. It towers up to the heavens and blocks the stars, a shadowing hulk of black iron whose wide girth spreads out over you in deathly menace.

Suddenly, you begin seeing sparks, as the side of the behemoth scratches against your small boat. You rush to grab combustible items to keep them away from the sparks. The grinding and sparks grow worse and worse. Fires are starting in your boat. You put one out and another begins. Soon there are too many to extinguish. You bail water into the boat to put out the fires, but the water starts to sink the boat. Then, in horror, you realize that the great ship is coated with some sort of sticky tar, an oily gunk that it has smeared along the edge of your boat. No amount of water will stop it from burning.

You grow weary in fighting to survive. You cannot sleep. You cannot escape. You can only watch the flames consume your boat and your belongings.

While this analogy is not perfect, it gives a sense of the magnitude of "the Dark Intruder." It spreads out over a colossal volume of space. The central planetary body itself may not be much bigger than the earth, although it may be denser, being composed almost totally of iron and other heavy elements. But, it is the breathtakingly large debris field that makes this monster so horrifying.

From stem to stern, this titan of the heavens is as broad as the inner solar system. The cloud of debris has generated clusters of fireballs that bombard the earth throughout its entire orbit around the sun, a volume of space nearly 200-million miles in diameter.

It takes a lot of gravity to hold a sea of rocks together for a 5,125-year circuit of the heavens (in 365.24-day years). Typhon cannot get much closer to the sun than earth's orbit, or this debris could be ripped away from its grasp. Yet the broad swirling mass of debris has remained gravitationally 'loyal' to the Dark Intruder for many, many orbits.

How many times has Typhon dragged all this debris past the earth?

One clue is the timing of the orbit. It turns out that eight such orbits total almost 41,003 years. Among climatologists, 41,000 years is famous as one of the Milankovitch cycles, the one related to the bobbing up and down of the earth's axis, that is, the degree of its tilt. Milutin Milankovitch was a Serbian mathematician at the University of Belgrade who calculated various long-term motions of the earth.

Milutin Milankovitch

No one has discovered the astronomical cause that lies behind the 41,000-year Milankovitch axis-tilt cycle. (Graham 2000)

But, having a massive object barreling past the earth (in a super-close approach), every 41,003 years (one out of every eight Typhon orbits) could explain that axis-tilt cycle.

This leads to an additional observation: If Typhon's close encounters affect the tilt of the earth's axis, then it makes sense that the times of maximum tilt, the solstices, are the times when Typhon comes by: Typhon causes the solstices by causing the tilt.

Not only does Milankovitch's axis-tilt cycle of 41,000 years sync with Typhon's orbit, but so does the long axis of the ellipse of the earth's own orbit. Specifically, the point at which the earth is farthest from the sun is reached on July 2, which is within two weeks of the time when the core of Typhon appears to make its pass by the earth.

There's still more: The point when the earth and sun are lined up with the center of our Milky Way galaxy is also within two weeks of Typhon's apparent passage.

Could all these alignments be due to chance?

Hardly likely... This fourteen-day period around the summer solstice is $1/26^{th}$ of a year. Thus, there is only one chance in 26 that any given annual astronomical event will occur during this specific period of the year.

For any two of them to occur during the same fourteen-day span is 26 x 26 (26^2), or one chance in 676. But for any three events, the odds rise to one in 17,576 (26 x 676).

And, there is but one chance in 456,976 for all four to happen with this specific fourteen-day window (earth's axis-tilt, solstice, galactic alignment, and greatest elongation from the sun = 26 x 17,576 = 456,976). With these odds, it is extremely unlikely all four relationships occur together by chance alone.

Finally, what are the odds that Typhon's debris field should come by in the very heart of this same fourteen-day window? Once again, we multiply 26 times the 456,976:1 odds obtained for the other four alignments: The result of this calculation is that there is but one chance in 11.9 million that all these events occurring together is a random, unrelated circumstance. Therefore, it is almost certain that Typhon is related to most, if not all, of the astronomical oddities that occur in this 14-day period centered on late June.

If that is the case, then it means that Typhon is probably responsible for pulling the earth outward, away from the sun, stretching the long axis of our orbit and tipping our axis. That would require Typhon to be a very massive cosmic body.

All these gravitational effects are evidence Typhon must be an impressive object, a planetary body that may be as heavy as the earth itself.

What, then, is Typhon's origin?

The debris field suggests that there had been an explosion or a collision of some kind. It happens that we may be able to date this violent event.

Back in 1976, the Milankovitch cycles were compared with long-term climate cycles using ice-core data from Antarctica. In a famous cover article that year in the journal Science, the publication of the American Association for the  Advancement of Science, Dr. James D. Hays (pictured to the right) of Columbia University's famed Lamont-Doherty Geophysical Observatory, and a team of leading climatologists from his and other institutions, showed that the various Milankovitch cycles--including what we argue was Typhon's 41,000-year cycle--matched up perfectly with Antarctic core samples showing the ice age cycles for the last 400,000+ years. Dr. Hays wrote that the cycles fit together "like a key in a lock." (Hays, Imbre and Shackleton 1976)

When I discussed these findings with Dr. Hays at the time, I had no idea that there was such a thing as Typhon, but I did inquire if there were signs of a global flooding event in the data. The most recent such event that can be clearly identified was the sudden 375-foot rise in sea-level about 11,600 years ago that Dr. Cesare Emiliani had already found three years earlier, in 1973. We have seen that this event, which Emiliani linked to the sinking of Atlantis in a

January 1974 report in the New York Times, can be traced to a unique confluence of solar and planetary cycles not directly related to Typhon. (Muck 1978, foreword)

What I eventually came to realize, however, was that the rest of the climate fluctuations that Hays and his team were looking at, the ones involving the Milankovitch cycles, may be mostly caused by Typhon's periodic disturbances of the earth.

Put simply, it appears that Typhon caused the ice ages.

The ice ages are known collectively as the Pleistocene. This era of ice ages began 3.2-million years ago, rather suddenly. The ice ages span the era of explosions by the super-volcano now located underneath Yellowstone National Park. One theory says this volcano is due to a "hotspot" in the mantle that the crust of the earth moves over. The earth's outer crust can slide rather freely over the mantle because the mantle is a molten hot liquid of melt rock. Such slippery motion of the crust could occur during a close-encounter with a massive object like Typhon, thereby setting off volcanic activity.

If one ignores the Milankovitch cycles for a moment, it would be easy to accuse the Yellowstone volcano of causing ice ages by spewing its megatons of ash into the atmosphere. But volcanoes also produce prodigious amounts of carbon dioxide, which is supposed to warm the earth (although some dispute this). Recent such volcanic eruptions of ash and carbon dioxide seem not to move global temperature much one way or the other for long. (Perkins 2006)

Granted, Yellowstone's eruptions are greater than any we see today. The problem with attributing the ice ages to them is that there have been only a few such mega-explosions. There have been more ice-age cycles than Yellowstone alone can explain.

That's where Milankovitch and Typhon enter the equation. Hays et al showed changes in the way the earth receives radiation from the sun (which Milankovitch's cycles explain readily) nicely match the coming and going of the ice ages. They show that when the orbit of the earth and its tilt cooled the continents sufficiently, there was more ice.

But, something is missing from the calculation: Dust. We now know that without a big increase in atmospheric dust, there would not be enough nuclei for condensation to create the precipitation that forms glaciers, and certainly not the three-mile-high variety that weighed down North America at the peak of the last ice age.

Moreover, this dust must be supplied very suddenly, for the record shows that the ice fell from the sky with alarming speed and built up almost instantly. If one views the ice data over the past 3.2-million years, what is most striking is that the sea-level plunges and the ice piles up, essentially overnight. Then, it gradually melts for thousands of years. (Morris and Johnson 2012, chart on p. 13)

In order for all that ice to be generated so rapidly, something must quickly inject massive amounts of dust into the atmosphere. And, this rapid injection of dust must occur while the sea is heated up so much that it overpowers the earth's cool orbital orientation (the Milankovitch cycles). The sea must suddenly get hotter in order to supply the energy to evaporate up the tremendous quantities of sea water needed to create gigantic glaciers so quickly. As just noted, the Milankovitch cycles should be cooling the earth and (in theory) reducing solar heating of the seas at the very moment when the sea levels are suddenly plunging (= evaporating) in order to produce the rapid ice buildup.

How can the seas be hot and cold at the same time?

Typhon solves the mystery. Its huge debris cloud includes both a huge amount of dust... and also large rocks. As Typhon draws closer, the number and size of objects in its debris field hitting the earth increases. During Typhon's close approaches, its larger orbiting bodies, burning with fire, would strike mostly in the sea (earth is 75% ocean). The sea would boil with the heat, suddenly evaporating up stupendous volumes of water. The cooling effects of the Milankovitch cycles would then ensure that the precipitation formed from the influx of Typhon's dust would later fall as snow and ice, not rain.

What we see here is a cosmic machine. Typhon comes by on its inevitable cycle and showers the earth with dust. Sometimes it also pummels us with great meteorites and comets. If they hit glaciers, they melt them, as happened 15,000 years ago.

But, if they fall into the sea, they evaporate up a lot of water, generating clouds for a sudden deluge of precipitation. If the Milankovitch cycles place the earth in a cooling orientation, that precipitation will form huge glaciers, a new ice age. But if the cycles have placed the earth in a warming orientation, we get deluges of rain and a new era of agriculture and population growth, as we have been enjoying for the last 11,600 years.

Without Typhon, the ice ages are a mystery. With Typhon, they make sense.

So, the comings and goings of the ice ages that began 3.2-million years ago can be easily explained by the way Typhon has interacted with the earth during that time. But why did all this abruptly begin happening 3.2-million years ago?

We can only infer from the sudden change in earth's climate at that time that our orbital relationship with Typhon had also undergone a sudden change at that time.

That is, Typhon's orbit had been changed in order to bring it near the earth every 5,000 or so years, as now happens. What could have changed the orbit of an object as big as Typhon? Could it have hit something? That would explain the bowl-shaped gouge in it and the enormous debris-field that accompanies Typhon.

We are nearing the solution of many mysteries. About 3.2-million years ago, as it orbited out in the solar system, Typhon made a close approach to another object. It was big enough to do severe damage to Typhon, wrenching out a gigantic gash in Typhon's iron bulk, much like an iceberg gouging a hole in an ill-fated ocean liner. Icebergs are far larger than the ships they damage. Some 3.2-million years ago, out in the dark ocean of space, Typhon encountered an ice-cold Behemoth far larger than itself.

What did Typhon hit? What huge ice-planet lay in its former path? Is there any sign of a gigantic derelict ball of damaged ice drifting through space in that region?

Yes. The planet Uranus is such an object. Sometime, only a few million years ago, Uranus was struck by an earth-sized object.

(Dickinson 1999, p. 50) It tipped Uranus over onto its side, like a beached whale, and giant Uranus, fifteen times the mass of the earth, has lain on its side ever since. (Dickinson 1999, p. 50)

For something the size of the earth to have accomplished such a great feat, it had to have been moving at extremely high velocity. That is no ordinary planet. The object that struck Uranus had to be approaching it as fast as a comet. It would have been a planet-sized body in a comet-like orbit. That certainly describes Typhon.

If we could peer down into Uranus' milky atmosphere, full of ice and carbon dioxide, we would see a great wound on its surface where it had been stuck a glancing blow by some high-velocity Intruder. That errant invader must have bored through the icy atmosphere, searing it with fire, as it burned its way into Uranus like a bullet, grazing its frozen heart.

That cosmic bullet had to have not only been gouged itself, but it must have gored a large mass of material out of the surface of Uranus. Of what is that surface composed?

Calculations of the contents of the atmosphere of Uranus and its tremendous pressure have produced the surprising result that it rains tiny nano-diamonds onto the surface of that planet. (Dickinson 1999, p. 48) The atmosphere crushes the carbon in its air into diamonds, which then drift down to the surface, building up a glassy accumulation of diamond dust over the course of billions of years.

That means that the object that hit Uranus must have picked up a pile of nano-diamonds, and this diamond dust should now be surrounding it in a large cloud of debris from the collision. Is this true of Typhon?

Yes, comets are a very significant part of the Typhon debris field. Studies of the make-up of comets reveal that their carbon is 10% diamond dust. (Carlisle 1995, p. 68) The origin of comets is much-debated. One mystery is how short-term comets are replenished. To explain the number of such comets requires that passing stars must frequently interfere with an assumed "Oort Cloud" of comets on the outer edge of the solar system. (Carlisle 1995, p. 66)

But, there is no proof that the Oort Cloud even exists, much less that other stars pass by often enough to keep comets plunging in to visit us. It is all theoretical, devised to explain the existence of comets on the unproven assumption that comets form out in deep space. But, some astronomers have disputed the Oort theory. Some comets may have been formed early on, but if so, that would not prevent others from being formed much more recently and violently. (van Flandern 1993, pp. 155-236)

Explaining why 10% of comet carbon is diamond dust has been another problem not satisfactorily resolved. Nano-diamonds, as we noted, require extreme pressures to form, something lacking in the wispy outer reaches of the solar system where the Oort Cloud is supposed to reside. (Cowen 2006)

On the other hand, Typhon-related comets should have nano-diamonds, if they are composed of icy material gouged out of Uranus' icy atmosphere and its nano-diamond surface. Typhon easily explains why our current comets have so many nano-diamonds, but also why there are so many such comets at the present time.

How long ago did Typhon hit Uranus?

Our assumption that this event led to the start of ice ages here on the earth needs support regarding timing. Did this event take place 3.2-million years ago, or far earlier in cosmic time?

Unlike the other three gas giants, Uranus has a rather uniform appearance, with few signs of the colorful banding that delights astronomical calendar-makers. No one is rushing to put out posters of Uranus. It has a milky kind of atmospheric boredom that may be due to murky turbidity left over from its impact event. That is, Typhon could have stirred up and "homogenized" the atmosphere of Uranus, destroying the banding of its former appearance. This milky appearance, then, implies that Uranus has been struck rather recently, as opposed to being hit early in solar system history.

But, there is another line of evidence about the timing of such a major planetary cataclysm in the outer solar system. Several astronomers have proposed that Pluto may have suffered from a major collision within the past ten-million years. It seems the orbit of Pluto's largest moon, Charon, has not yet circularized sufficiently, as would be expected if the collision that formed it had occurred more than ten-million years ago. (Highfield 2006)

Since this estimate, two rather small moonlets have also been discovered orbiting Pluto, but much farther out than Charon. Astronomer Dr. Robin Canup assumed formation of the three moons was caused by large-scale meteoric bombardment during the early solar system. (Highfield 2006) She theorized that synchronicities of the three moons meant that they had all formed when a single "massive object

Robin Canup
Planetary Science Directorate

walloped Pluto" at that time. This object had been estimated to have been 250 kilometers in diameter by earlier researchers.

Whenever this collision occurred, it changed Pluto's orbit, currently part of the time inside the orbit of Neptune, that is, on the Uranus side of Neptune. Had Pluto once orbited entirely inside Neptune and was stuck by an object coming from the direction of Uranus, it could have been jettisoned out into the more exterior orbit in which we now find it. Given the fact of the Uranus impact, such a scenario is fully credible.

On the other hand, if it had been hit by some unknown errant Kuiper Belt Object from outside of Neptune's orbit, then it would seem to have no relationship to the early inner solar system bombardment to which Dr. Canup credited the Pluto collision.

Moreover, at that early date, it is unlikely that big objects existed in sufficient numbers in the Kuiper Belt to cause such a major collision, even if one speculates that Neptune itself slowly moved outward, jostling the belt. (Kenyon 2004) Some think Neptune migrated from about halfway between where Uranus and Saturn currently orbit to a place 50% farther out from the sun, taking the Kuiper Belt outward with it. (Kenyon 2004)

Astronomer Dr. Tom van Flandern spent much of his career developing the idea that there had once been a Saturn-sized gas giant orbiting where the asteroid belt is most stable (about 250-million miles from the sun). (van Flandern 1993, 155-236) He first decided it had exploded 3.2-million years ago, creating most of the large asteroids and comets and indirectly initiating the ice-ages that began at that time. (van Flandern 1993, 155-236) He did note,

Dr. Tom van Flandern

however, that the events of 65-million years ago, when the dinosaurs died, also seemed to reflect the same kind of conditions, but he was unsure of the connection (van Flandern 1993, p. 235).

In his 1993 book, *Dark Matter, Missing Planets and New Comets*, Dr. van Flandern listed dozens of scientific observations about comets and asteroids pointing to a violent origin for many of them 3.2-million years ago. In particular, he emphasized that new comets, the ones that had never previously approached the sun, shared that 3.2-million-year age, proving a common origin at that time. Moreover, he noted that they uniformly began to lose matter and form tails of debris as they passed within the 250-million-mile proposed orbital distance where the gas giant had been. If these comets been formed at that distance, he felt that when they got closer than that, the stronger solar gravity was great enough to disrupt them. Later, he noted signs that the planet may have exploded earlier in time. (van Flandern 1993, p. 235)

In either case, Typhon has been orbiting through the outer solar system for millions of years. Therefore, it has repeatedly passed through the Kuiper Belt, and its gravity has affected many of the objects there. Is there any evidence for an earth-sized planet like Typhon passing through the Kuiper Belt on a frequent basis?

Yes, the Kuiper Belt shows "sculpting" by the passage of an earth-sized planet moving through it on a regular basis over a long period of time. (Trujillo 2003, pp. 426-429) This earth-sized planet would have been in a comet-like orbit, jostling around the much smaller Kuiper Belt Objects; this may have been occurring for perhaps billions of years, gradually creating the arrangement we now find there. (Trujillo 2003, pp. 426-429)

We need not speculate on what this errant earth-sized object might be. There are no known Kuiper Belt objects anywhere near that large. The only such earth-sized object known to have a comet-like orbit is the one that knocked Uranus onto its side, that is, the object I have identified with Typhon.

We can see that Typhon must have been a long-time resident of its chilly living-room in the Kuiper Belt, and it would have kept itself busy, re-arranging the icy furniture out there for many millions (or billions) of years.

That would mean that Typhon has been a long-term feature of the solar system, perhaps going back to its early days. How did this big iron ball get formed?

When we survey the inner structures of most planets, we find that an iron core is normally a small part of the total volume. Typhon's high density implies that it was once part of a far larger planet. Is there any evidence that such a large planet was ejected out of the major planetary family and hurled outward toward the Kuiper Belt?

Yes, computer simulations of the early solar system require such a scenario. (Grant 2012) It was impossible to reconstruct the development of the solar system unless there had been another giant planet (at least as large as Uranus) in the original solar system that had been ejected by coming too close to Jupiter. (Grant 2012) Such a planet ought to be "out there" somewhere. What happened to it? We need not wonder. Typhon fits the description. But we need to ponder some other curious facts:

The asteroid belt has 1/1,000th of its original matter. (Binzel 2001) It had enough solid material to make c. six earths: as much as the mass of Saturn's core. (Binzel 2001) Thus, material for a Saturn-like planet had once orbited near or in the asteroid belt. Where did it go?

Studies of Kuiper Belt Objects (KBOs) reveal that they originated much closer to the sun. If KBOs began as part of a gas-giant in the asteroid belt, which is an unstable region for a planet to form due to Jupiter's gravity, then that Saturnine gas-giant would have been broken up and ejected by Jupiter. (Binzel 2001)

But, computer simulation found only one such large planetary ejection. We must observe, then, if it happened only once, there could not have been three such separate events. The three events, for which only one common explanation is possible, are:

1. A gas-giant planet was ejected by Jupiter into a highly elliptical orbit, based on computer reconstruction of the early solar system.

2. The asteroid belt's large planet-sized mass (= 6 earths = the core of a Saturn-sized gas giant) was ejected from the asteroid belt by Jupiter.

3. And a large "missing" planet, in a highly elliptical orbit, hit Uranus after it had sculpted the Kuiper Belt into the structure we see today.

There was only one such planet. It would have tried to form in the asteroid belt region of the early system, was ripped apart, and much of its remains were ejected into the outer part of the solar system to form and sculpt the Kuiper Belt.

Estimates of the total mass of material now in the Kuiper Belt are obviously going to be subject to change as new objects are found. One estimate in 2003 was that its total mass (for its 30,000 largest objects over 100 kilometers in diameter) would be 100 times the current mass of the asteroid belt, or a bit smaller than the earth. (Trujillo 2003, p. 424) If we add in scattered dust and tinier rocks, the mass might approach that of the earth.

Typhon plus the Kuiper Belt Object's estimated mass leaves us with about four-earth's worth of missing mass from the 6-earth masses of debris Jupiter ejected from the asteroid belt.

Obviously, not all of the exploded debris of the doomed planet is still orbiting around the solar system. Some of it must have hit the various moons and planets. Perhaps one-earth mass of this debris has bombarded solar system bodies or become moons, but that still would leave half of the original six-earth masses unaccounted for.

There is this missing mass? What happened to the iron core of the ejected planet?

When Typhon's parent planet became unstable, its spin-rate and disruption by Jupiter caused a catastrophic explosion, one big enough to divide the core into two halves, each half ejected into the outer solar system.

Based on the mass of the planet, the core should have been much larger than Typhon alone can account for. So, we must look for a second core portion, one cast off in the opposite direction, in the violent dissolution of this giant Saturn-sized planet.

How might this have happened?

A gas-giant planet is mostly cold or frozen gas. If the sun heated it up, the gas would begin to dissipate. As the planet lost gas, it would shrink. As it became smaller, its rate of rotation would increase. The gas-giant planet Saturn rotates in only ten hours. The other Saturn-sized planet on the inner side of Jupiter's orbit would have been spinning far faster than Saturn, as its gas evaporated. The faster spin would eject still more gas.

This rapid rotation escalated geometrically. As noted, giant Saturn currently spins every ten hours, yet still retains most of its original frozen gas. Typhon's parent, on the other hand, lost much or all of its frozen mantle. So, it had to spin many times faster than Saturn now spins, faster than once an hour.

The internal centrifugal forces created by such super-fast rotational rates would begin ripping the planet apart. At the outer edge, rotation at the equator would exceed gravitational escape velocity. (van Flandern 1993) The planet could begin casting off debris from the outer surface, which would further accelerate the spin.

The doomed planet also felt the continual tug of Jupiter. The orbits of the two large planets became attracted to one another. Having the gravitational effects of Jupiter passing ever closer steadily escalated the pressures on the fragmenting planet's structure. Eventually the spin-rate escalated so much the core split catastrophically.

It was at this point that chaos took over. The core was driven out through what was left of the fractured hulk. Debris hurtled outward. The entire solar system would have been pelted by the initial debris cloud because it was flung so far afield.

Such a bombardment is known to have happened very early in the history of the solar system, and it was about that same time that a Mars-sized object hit the earth and formed the moon. I previously argued that this Mars-sized object became molten after it collided with the earth and coalesced into the planet we now call Mercury. The impact with earth should have reduced Mercury's orbit, which means Mercury had previously orbited farther out, closer to the asteroid belt. An exploding planet there would explain why proto-Mercury had been pushed into the path of the earth.

The pieces are now falling into place. The debris from the explosion bombarded proto-Mercury and this degraded its orbit closer to the sun. Richard C. Hoagland has suggested that the other Mars-sized planet, Mars itself, had originally been a moon of the exploded planet. If so, we may speculate that this doomed planet had two such moons, one of which later became Mercury after it hit the earth and formed our own moon.

All inner planets show the effects of this early bombardment from the shattered gas-giant. Still, we are left with a puzzle: What became of the other half of the core?

The first astronomer to calculate the effects of an explosion in the asteroid belt was the late Dr. Tom van Flandern, formerly of the U.S. Naval Observatory. (Hancock 1998, p. 40) He found a periodicity in the orbits of major asteroids and comets that suggested that many of them had been clustered together into a fairly tight region of space in the asteroid belt. But as we noted, he was troubled by indications the explosion could have been taken place when the dinosaurs died, presumably about 65-million years ago. (van Flandern 1993, p. 235)

However, there is a phenomenon in orbital mechanics where objects continually return to their starting-point. Periodically, large numbers of such dispersed objects will return at about the same time. This clumping will repeat at about the same cycle over long periods. So, 65-million years ago need not be the original planet-explosion event.

Thus, if the 65-million year ago clumping was just the latest in a much older cycle, we should find evidence of it repeating over a longer time-frame. And, indeed, that is exactly what the record shows... a major extinction event, 65-million years before the dinosaur extinction. And yet another about 65-million years before that, and so on. The earliest hint of this cycle goes back about 585-million years, nearly 600-million years ago. Before that, we don't have extinctions of calcified fossil creatures, because such bony structures had only just then begun to appear on the earth.

Not coincidentally, nearly 600-million years ago, there was a major bombardment of Venus, the moon, and other parts of the inner solar system. Something extraordinary had happened. It was the worst bombardment since the early days of the solar system. What had taken place 600-million years ago?

Imagine the core of a planet that had exploded approximately 4-billion years ago and that had divided primarily into two fairly similar chunks of iron and other heavy elements, each jettisoned out into space on a great looping orbit, destined by orbital mechanics eventually to "return to the scene of the crime" on the same day.

Now imagine that this fateful day had come *c.* 600-million years ago and that the two core fragments had returned at exactly the same moment to the exact spot where they had burst asunder billions of years before.

Had this happened, the two hunks of core material would have collided with an abominable release of energy. This clash of titans, each of which then outweighed the earth, would have devastated everything within the orbit of Jupiter and would have sent projectiles out beyond Neptune. And that is exactly what happened.

What survived were two reduced globs of molten iron, each one cast out into the dark depths of space. One of these became Typhon.

The other went off in another direction, hurled into the heights of heaven. This missing mass has cooled into an iron sphere. Like Typhon, it carries along a cloud of debris. Unlike Typhon, it never hit Uranus. It retains all its original momentum and mass. So, when it returns on its long orbit, it brings horrendous devastation upon the earth.

As we may infer, Typhon returns occur more often than just every 65,000,000 years. But, it is every 65,000,000 years that it comes closest to the earth. On most of its other orbits, the earth is well away from the visitor. But upon these fatal meetings, the earth's continents are ripped asunder in a paroxysm of mega-death and mass-extinction.

We can now examine the demise of the dinosaurs for an insight into Typhon's deadlier twin. Previously we saw that several objects impacted the earth 65-million years ago, one in the Yucatan Peninsula, one in Manson, Iowa, and three more in Canada and Siberia. It has been assumed that the Yucatan Peninsula impact was both the first and biggest event, no doubt because it had been the first impact site related to the dinosaur extinction.

However, there is no requirement that the Yucatan Peninsula impact could have come in any specific sequence. Could the destruction of the dinosaurs have come about at the climax of a global assault by multiple impacts and calamitous upheavals?

Scientists have found that, at the very same moment the dinosaurs died, India underwent one of the greatest outpourings of lava in the history of the planet. Known as the Deccan Traps, these lava flows erupted on the opposite side of the earth from the great Yucatan impact. (Hancock 1998, pp. 44, 193)

This phenomenon of great lava outpourings on the opposite side of a planet from a major impact is not unique to the earth. The same thing is found on both Mercury and Mars, and should also have occurred on Pluto, which was also struck by a proportionately huge impact that formed its large moon Charon.

But 65-million years ago, when lava poured out onto the Indian plateau, India was not anywhere near where we see it today. At that time, India was farther south than the island of Madagascar is now, about even with the southern end of Africa. On the floor of the Indian Ocean, we can still plainly see the tracks of India as it slid rapidly northward, in a perfectly straight line, until it collided with Asia to form the Himalayan mountains (National Geographic Society 1967).

What is all the more amazing is that this long-known movement of India must have been so rapid that it never veered from its straight track and it hit Asia with such force that it moved the sea-floor up to the highest elevations on earth.

Just how fast was India moving?

Well, if we examine the sea-floor evidence closely, it shows a huge undersea lava flow right out of the depths of the ocean. (National Geographic Society 1967) India was sitting directly over it at the time, and we must assume that India's Deccan lavas came from these undersea vents.

It appears that Typhon's twin, call it Typhon II, had come by the earth 65-million years ago in a passage so close that it nearly sterilized the planet. It cast several large chunks of debris at the earth, one of which was big enough to cause a huge eruption of lava on the far side of the planet. As we noted, India was then a large island in the south Indian Ocean sitting on top of the erupting lava.

A vast layer of liquid lava spread out beneath India. The entire island of India was lifted up bodily and began to slide northward on its lava-slick belly. We can see the sea-floor streaks obliterated where the lava vents continued to gush after India slid off to the north, (National Geographic Society 1967)

The problem is, India was sliding uphill. The sea floor where the lava erupted was lower than the sea floor to its north. (National Geographic Society 1967) What moved India uphill?

If Typhon II was passing northward above the earth just over India, it could have become gravitationally locked onto India. (van Flandern 1993, pp. 148-153) In that case, it would have been able to carry India along with it, dragging the subcontinent northward through the ocean, until it collided with Asia, which then broke Typhon II's grip on India.

If this were to have happened, India's bulk plowing through the sea would have produced tidal waves several miles high, easily able to deposit the fish fossils found atop the Himalayan Mountains today. But the tidal waves of salt-water from the Indian Ocean would also have salted the soils of East Africa, Arabia, southern Mesopotamia,

Persia, and Western Australia, all of which have become deserts.

It may even be that the eruption of the Indian Ocean lava flows was occasioned only by Typhon II, and not by the Yucatan impact at all. In this case, the Yucatan might even have been the last site impacted, struck as Typhon II was moving away.

Looking at the events globally, Typhon II swept in over the south Indian Ocean, erupting the mid-ocean lava vents, greasing the bottom of India with what might as well have been billions of ball-bearings, lifting it up and dragging India through the ocean, flooding the entire surrounding basin with great quantities of salt water.

And there is further evidence for this scenario. The Indus and Ganges river deltas that annually dump their muddy waters into the Indian Ocean have piled up great fans of mud on the sea floor. There is well over a vertical mile of these sediments, and they spread out from where India is now located.

In other words, it appears that India was swept northward into Asia first, and only then did these great rivers begin piling up their effluent on the floor of the ocean. There is no sign of accumulated effluent on the floor of the Indian Ocean back to the south where the lava vents are. Yet, the streak marks where India got dragged northward show over a thousand miles of effluent-free sea-floor that India had rapidly passed over.

This lack of sea-floor effluent implies that not a single major rain-storm washed India as it moved over a thousand miles through the ocean. No known mechanism can explain such a rapid movement...unless a cosmic intruder had lifted India up and had carried it through the sea in a few hours, that is, at supersonic speed.

The mid-ocean rift that Typhon II erupted is one that already existed and extended around Africa and up the Atlantic to Iceland and into the Arctic. We know that this rift was created far earlier than 65,000,000 years ago. It zigzags up the Atlantic in exactly the same shape as the coasts of Africa and the Americas on either side of it. Therefore, that zigzag shape pre-existed the splitting apart of the continents.

The continents began to split apart roughly 195,000,000 years ago, just about two cycles of 65,000,000 years before the dinosaurs perished. Typhon II might have been the trigger that began that great continental fracture that created the Atlantic Ocean.

Yet today, nearly 200,000,000 years later, or almost the same amount of time it takes us to orbit the entire Milky Way galaxy, that great mid-ocean rift retains exactly the same ancient zigzag shape it had before the continents split apart.

This strongly suggests that whatever created that peculiar shape inflicted more damage to the earth than even Typhon II has in all its repeated visits.

If one looks carefully at the sea floor, there is a fainter mid-ocean ridge below the fresher lava-flow gash that erupted below India. This is clearly an even older rupture in the sea floor, and from the look of it (when compared to the 65,000,000 year ago deposits), it had been there for many hundreds of millions, if not billions of years prior to the events that generated the Atlantic Ocean over the last 200,000,000 years.

We said before that Mercury, when we believe it was a Mars-sized planet, had gouged the moon out of the earth's hide some 4-billion years ago. That a Mars-sized planet could have hit the earth at that time had been proposed as early as 1990, by a Harvard astronomer named Willy Benz. (Benz 1991)

It is now generally accepted that this Mars-sized object hit the earth. However, the earth may been hit, not once, but twice. One computer simulation found that the moon rock evidence required a first impact that was only a glancing blow. At some subsequent time, the earth was hit head-on, and the moon created as a result. (Lauber 1993, p. 23-25) It was Van Flandern's belief that the moon had emerged out of the Pacific Ocean basin, but he thought it was tidally-ejected, not due to an impact. (van Flandern 1993, p. xxxiii)

So, the two greatest impact events in earth history would seem to be two hits we received when the moon was created. The first was less damaging, but certainly would have caused dramatic upheavals. The subsequent event would have been by far the worst in our history.

These two events could have left behind the two ocean rifts, one older and much fainter in the Pacific and the other, in the

Atlantic and Indian oceans, that is clearly newer and much more jagged and pronounced.

The lack of change in the zigzag shape of the rifts over vast ages of time implies that they got their jagged shape not long after the impacts with the earth 4-billion years ago. The jagged sideways fractures may be the direct result of our brand-new moon, which was then orbiting just above the surface, creating enormous earth tides.

One of the peculiar aspects of the first rift, the faint one in the Pacific Ocean, is that it is off-center. The appearance is that of a rift that the continents have slid over, as if the rift were no big barrier. This is not the case with the second, younger Atlantic rift.

Also, the Pacific rift runs up the Gulf of California, between Baja and Mexico and then dives under North America. The left side of the rift appears to be the Sierra Madre and Cascade ranges. The right side may be the Rockies. So, like Sodom and Gomorrah before them, Las Vegas and Salt Lake City sit over an oceanic rift's huge lava caldron.

However, there is a faint trace that emerges as a faded remnant that ends in the midst of the Arctic Ocean basin. It appears that before North America was pushed over it by the newer Atlantic rift, the faded Pacific rift had once ended in the Arctic.

But, when the Atlantic rift shoved North America over the ancient Pacific rift, it put enough pressure on the dormant rift to revive its lava flows. Think of it like an old wound that had mostly healed, but then was ripped open again after being scraped raw by North America. The once-dormant portion of the Pacific rift began to hemorrhage lava via new volcanic activity under the continent. Those volcanoes are still active today.

The other end of the Pacific rift is mostly obliterated under the Indian Ocean rift, but hints of it can be found, headed generally west toward Africa. There it may have formed the valley of Lake Victoria, at the head of the Nile.

Does this help us find the locations of the early impacts 4,000,000,000 years ago?

Beneath the surface of the earth where the impacts occurred would have been lumps of the impactor's crust and mantle. These submerged lumps are called "mascons," and can be detected by their effects upon the earth's gravitational profile. It seems the equator of the earth has a "bulge," but there is also a bulge south of the equator. These two bulges stretch pretty much around the globe. If they are the two impact mascons, then it would seem the earth and the crust of the impacting object (presumably proto-Mercury) were turned into a liquid molten mass that caused the mascon material to be stretched around the earth like taffy. Dr. Van Flandern, as we mentioned, along with several other scientists, decided the major site of the event is now in the area of the Pacific basin.

There is one clue that might reveal the difference between the two impacts. In the South Atlantic, west of the tip of Africa, there is a curious geo-magnetic anomaly. It seems to be the remnant of an alternate South Pole and is located about 45-degrees south of the equator. Thus, a discrete South Pole has resided in two specific places on the earth. The implied alternate North Pole would have been in the North Pacific Ocean, south of Alaska. The mechanism that created these poles is disputed by scientists.

Nevertheless, the fact that these poles have no intermediate location testifies to the sudden and violent displacement of the poles from one location to the other. We can infer from this that the two pole locations are a reflection of the earth's orientations at the times of the two primordial impacts. The scientific community generally agrees that what hit the earth was a planet that was orbiting in the ecliptic plane as planets now do. From this fact, we may deduce it hit the earth somewhere near its equator of that time.

The trouble is, a South Pacific impact could be compatible with both the current poles and the other alternate poles, depending upon the actual degree of tilt of the axis at the time. In either case, the data are all consistent with each other.

There is at least one place on the earth where a giant upwelling of lava is currently flowing up through the mantle in a way that might be consistent with a planet-sized wound in the upper mantle. In the March 2, 1995, issue of the prestigious British journal *Nature*, "a massive upwelling" of lava from near the base of the upper mantle was reported by Kaj Hoernle of GEOMAR Research Center in Kiel, Germany, Yu-Shen Zhang of UC, Santa Cruz, and David Graham of Oregon State University. (Hoernie, Zang and Graham 1995)

Although some find their results 'suspicious' because these results contradict previous theories that such a mass-migration of lava ought not be happening, the upwelling cuts through 45-degrees north latitude, exactly opposite the South Pacific Ocean. (Hoernie, Zang and Graham 1995) Not only that, but that area in the South Pacific is close to the mid point of the Pacific rift itself. If that rift is the bleeding out of the lava from the far side of the planet from where an impact had occurred, then this makes sense.

There is still more. The newer Atlantic rift deviates dramatically westward around this upwelling site, creating the westward extension of the African continent at this exact location. It suggests that the upwelling forced the Atlantic rift to shift sharply westward to go around the upwelling region. The Atlantic rift then returns to its original "track" as it moves north toward the Arctic.

As a consequence, the very shapes of America, Europe and Africa would be the result of this deep upwelling of upper mantle lava that may designate the site of the first of the two impact wounds the earth received when the moon was formed over 4,000,000,000 years ago.

The second impact site depends upon how one interprets the Atlantic rift. It is hard to find the center of that rift because of the massive eruptions occurring 65,000,000 years ago along its Indian Ocean extension. The location opposite those eruptions is in the Pacific Northwest, roughly under the super volcano at Yellowstone.

But, if one looks for the rift center in the Atlantic, it is near the same site as the South Atlantic magnetic anomaly that designates where the alternate South Pole has been located. And its opposite location is, of course, near the alternate North Pole in the heart of the Pacific "Ring of Fire" due east of northern Japan. This region of the Pacific is often compared to the Bermuda Triangle, because it has had so many strange disappearances and magnetic anomalies. Undersea, it is a gnarled mass of giant volcanoes, deep ocean trenches, and vast lava flows. Perhaps that was the true birthplace of the moon.

The one thing that does emerge from all this is that the mantle of the earth should have a pair of lumps in it that are about 45° apart. Therefore, if Typhon or its twin were to make a really close approach, it could gravitationally "lock-on" to the outer crust and the upper mantle and slide or twist them around. But, the two lumps are like door-stops. They limit how far the crust can be made to slip around the mantle.

Such shifting of the crust relative to the axis of rotation might be reflected in the 41,000-year Milankovitch cycle of the degree of axis tilt. Anything that would cause one to occur could also result in the other happening. Since we know the axis has in fact been subject to tilting on a Typhon-related 41,000-year cycle, it is entirely possible that the earth got shifted back and forth between the two poles on a fairly frequent basis. So, it's conceivable that close passages caused a 45-degree crustal-shift when Typhon comes by.

Keep in mind that what we are describing is not a 45-degree tilt of the axis of rotation. That may have moved only a couple of degrees. The major shift is the slippage of the outer surface of the earth over its inner mantle and core, which should continue to rotate fairly close to the way they did before.

Is there any historical record that describes such a shift in the orientation of the surface of the earth?

Yes. Immanuel Velikovsky, in his 1950 book, *Worlds in Collision*, presented dozens of scientific, archaeological, and historical indications showing that the earth had a different orientation than it now has. (Velikovsky 1950, 1967, pp. 37-52; 118-137; 316-332; etc.) But the most illuminating reports came from ancient Egypt, where the Greek historian Herodotus reported the following account by the priests when he visited Egypt *c.* 450 B.C.:

> Four times in this period (since Egypt had become a nation), so they told me, the sun rose contrary to

its custom: (Herodotus interpreted the priests to mean that) twice (that is, in two epochs of Egyptian history) the sun (was seen in a position where it) was arising where it now sets; and twice, it was setting where it now rises. (Velikovsky 1950, 1967, citing Herodotus, II, 142)

Statements like these have earned Herodotus a rather poor reputation among modern historians. Not even our Typhon thesis can explain such circumstances. As it happens, we need not depend solely upon Herodotus. Pomponius Mela (a Latin geographer of the first century and a contemporary of Paul) said:

The Egyptians pride themselves about being the most ancient people in the world. In their official records... one can read that during the time that they have been in existence, the course of the stars has changed direction four times, and twice, the sun was (for a period) setting in the part of the sky where it is rising today." (Velikovsky 1950, 1967, citing Pomponius Mela, *De situ orbis*, i. 9. 8)

Note that Pomponius cites written records, not the spoken words of priests, and he also mentions "the stars" changing their direction four times, not just the sun. It appears that he obtained his information from the written records that the priests had used to relay their account to Herodotus. So, Pomponius seems to be a more direct source, albeit later than Herodotus. In Pomponius' version, the locations of rising and setting are more generalized. And, he indicates that the stars at night were also not in their accustomed motion.

We may not appreciate the significance of this today, for in his time it was not realized that the sun and the stars would both have to have changed direction together. Therefore, this account was most likely derived from actual observation, not merely from deduction or theory.

We mentioned previously that Egypt had been in existence for about 18,000 or so years, by their own reckoning. During that period, Typhon had come by three times. This would create four separate eras of Typhon, if one includes the era before the Egyptians would have first recorded Typhon coming by, that is, before c. 15,000 years ago.

Since we know the current orientation of the sun, rising from the east, we may reckon this orientation with the fourth era of Egypt. In the preceding era, between Typhon's visits 5,000 and 10,000 years ago, the sun supposedly rose in the west. Between Typhon's visits 15,000 and 10,000 years ago (the age of Atlantis and the Deluge), the sun would have risen where it now does, in the east. And finally, in the oldest epoch of Egypt, before Typhon's visit 15,000 years ago, it would (in theory) have been rising in the west.

Now I doubt that this was literally a 180° shift. Instead, I believe Typhon may have been responsible for a 45-degree shift (the shift of the crust between the two mantle 'door stops' that limit the range in which the crust can slide). But the sun at the Summer Solstice (when Typhon comes by) is actually rising well north of east, not due east. So, a 45-degree shift at that time of year could produce a very dramatic change in where the sun appeared to rise, and a very different perceived motion of the stars at night.

The original Egyptian record seems to have simply stated that "the sun and the stars rose twice in the history of the nation in alternating locations." This account can only have been composed after the last visit of Typhon 5,000 years ago, in c. 3,100 B.C. It must have been derived from earlier records or observations, but those cannot be what either Pomponius or the priests who spoke with Herodotus had referenced. We must infer that the term "alternating" was misinterpreted as "opposite," leading to a more extreme view of what the sun and stars had actually appeared to be doing in their motions.

Once we realize that the terms for "alternating" and "opposite" had been confused, we can see that the original Egyptian records were merely noting that the sun had gone back and forth between two specific apparent paths through their heavens, and that the stars had likewise alternated their motions.

All of this allows us to observe that Typhon seems to have become in an orbital synchronization with the earth's motion around the solar system. Its visits seem to have repeatedly occurred at the same time of the year, and thus, in the same part of our orbit. And that, after all, would support the emphasis upon the Summer Solstice timing and the way in which Typhon appears to have affected the earth's tilt and the elongation of our orbit at that point.

Unfortunately, that synchronization was based upon the old 360-day year, the one the Mayan count assumed, but which Mercury and Mars appear to have altered to 365.24 days since Typhon's last visit. We are not going to be where Typhon "expects" us to be. Accordingly, Typhon's visit this time could be far more dangerous...

Fireballs vs. Sightings

It is important to distinguish between the actual fireballs themselves and reported sightings of them, when discussing this subject. (All information in this appendix is from the website of the American Meteor Society: http://www.AMSmeteors.org.)

It is estimated that thousands of fireballs enter our atmosphere every day. Not many of these are observable, however, because half are obscured by daylight, and three-fourths of the remainder enter over unoccupied land or sea locations. Of those observed, few are reported. Of those reported, few contain the precise timing, azimuth, latitude and magnitude data needed to identify the object as a specific fireball. The upshot of this is that official "sightings" are reported out of only a small fraction of the total of actual fireballs entering each day. The data, therefore, represent a kind of "poll" of the fireball population, a sample that is as "scientific" as these limiting circumstances allow.

The data of the American Meteor Society (founded in 1911) are gathered from trained observers in the United States and Canada. Therefore, their reports filter out the reporting of other objects by amateurs. In this way, they can insure that nearly all of their data is derived from genuine fireball events.

To be an official fireball, the object must be a certain brightness, as if it had been seen through a clear sky directly overhead at sea-level. This is not an observation but is arrived at by a complex calculation. The brightness can rarely be measured perfectly but must be estimated by a trained observer in relation to other known objects in the heavens. This magnitude requirement for a fireball has to be adjusted for elevation above the horizon at moment of flame-out, as well as the altitude of the observer and atmospheric conditions at the sighting location.

Obviously, the general public will "report" sightings of bright fireball-like objects far more frequently than these trained observers. But, a call to a UFO reporting group will not end up in the A.M.S. data. They cannot use vague descriptions that do not allow them to identify the report with a specific object that can be distinguished from others reported around the same time. Fireballs, as we have noted, are coming in groups and clusters.

The A.M.S. data show a dramatic increase in fireballs (see chart at the beginning of this report). By July 2012, the total for the year was well past 1,000. In recent years, the totals in December would often be the highest for the year, due to the overall upward trend in fireballs. So, it is safe to estimate that 2012's total will surpass 2,000. Numbers of individual reported sightings are a few times higher than the number of fireballs because the same object (or cluster) event may be reported by several trained observers.

The A.M.S. does not feel comet debris reaches the ground intact, but only a few fireballs have had spectra taken to determine the materials they contain. Some are red, some green, and some fiery yellow-orange. Of these, only one kind may derive "directly" from Typhon. In any case, the total number of sightings by both trained observers and the public is increasing at a rate that shows a massive cluster of objects is now entering our planetary neighborhood.

Chapter Bibliography

Benz, Willy. 1991. "Collision Course." *The Wilson Quarterly*, Spring: 128-129.

Binzel, Richard P. 2001. "A New Century for Asteroids." *Sky and Telescope*, July: 44-51.

Carlisle, David Brez. 1995. *Dinosaurs, Diamonds, and Things from Outer Space*. Redwood City, California: Stanford University Press.

Cotter, Barbara, Lauren Fellers, and Sue McMillin. 2012. "WALDO CANYON FIRE: Day-by-day timeline - column of smoke west of Colorado Springs turns into deadly wind-blown monster." *The Colorado Springs Gazette*, July 5: 1. Accessed May 13, 2018. http://gazette.com/waldo-canyon-fire-day-by-day-timeline-column-of-smoke-west-of-colorado-springs-turns-into-deadly-wind-blown-monster/article/141248.

Cowen, Ron. 2006. "Comet Sampler: Fire meets Ice." *Science News*, March 25: 182.

Dickinson, Terrance. 1999. *The Universe and Beyond*. Buffalo, New York: Firefly Books.

Graham, Steve. 2000. *Earth Observatory: Where every day is Earth Day*. March 24. Accessed May 13, 2018. https://earthobservatory.nasa.gov/Features/Milankovitch/.

Grant, Andrew. 2012. "The Solar System's Lost Planet." *Discover Magazine*, May: p. 14-15. http://discovermagazine.com/2012/may/09-the-solar-systems-lost-planet.

Hancock, Graham. 1998. *The Mars Mystery*. New York: Crown Publishers.

Hays, J. D., John Imbre, and N. J. Shackleton. 1976. "Variations in the Earth's Orbit: Pacemaker of the Ice Ages." *Science*, December 10: 1121-11362.

Highfield, Roger. 2006. "Pluto's new moons give clue to massive collision in space." *The Telegraph*, February 23. https://www.telegraph.co.uk/news/worldnews/1511295/Plutos-new-moons-give-clue-to-massive-collision-in-space.html.

Hoernie, KAJ, Yu-Shen Zang, and David Graham. 1995. "Seismic and geochemical evidence for large-scale mantle upwelling beneath the eastern Atlantic and western and central Europe." *Nature*, March 2: 34.

Kenyon, Scott. 2004. "Cosmic Snowstorm." *Astronomy Magazine*, March: 29, 43-46.

Lauber, Patricia. 1993. *Journey to the Planets*. New York, New York: Crown Publishing.

Morris, John D., and James J.S. Johnson. 2012. "The Draining Floodwaters: Geologic Evidence Reflects the Genesis Text." *Acts and Facts*, January: 12-13. http://www.icr.org/i/pdf/af/af1201.pdf.

Muck, Otto Heinrich. 1978. *The Secret of Atlantis*. New York: Times Books.

National Geographic Society. 1967. "Indian Ocean Floor - Map." *National Geographic*, October.

Perkins, Sid. 2006. "Krakatoa stifled sea level rise for decades." *Science News*, February 18: 110.

Trujillo, Chadwick A. 2003. "Discovering the Edge of the Solar System: Recent Discoveries Suggest that Planets Larger than Pluto may exist in the outer reaches of our solar system." *American Scientist*, September-October: 424-431. http://www.jstor.org/stable/27858273.

van Flandern, Tom. 1993. *Dark Matter, Missing Planets and New Comets*. Berkeley, California: North Atlantic Books.

Velikovsky, Immanuel. 1950, 1967. *Worlds in Collision*. New York: Dell Books.

Chapter 8

To Cast Fire on the Earth

When and how will our "world" end? Scientific data can shed light on these questions. The Milankovitch Planetary Cycles match the climate-changes, extinctions and cultural breaks in geo-history. Those synchronized cycles indicate we are now due for yet another global-extinction event and ice-age. (Felix 1997, pp. 174-210)

For example, the 23,000-year cycle matches a magnetic reversal 23,000 years ago. (Felix 1997, p. 205) The latest ice-age began then. Following an extinction, mankind began weaving cloth, creating musical instruments, domesticating animals, and growing grain (Genesis 2:5, 15, 3:19, 21, 4:2, 17, 20-22, 5:29, 6:14). They milled flour, baked bread and built towns (Genesis 3:17-19, 4:2, 17). This led to trade and using numbers. Our first language began *c.* 20,000+ years ago. (Velasquez-Manoff 2007, citing *http://ehl.santafe.edu/intro1.htm*)

The mitochondrial DNA we all share is from one woman, a "genetic Eve," who is said to have lived less than 200,000 years ago and 23,000 years ago. The first date estimates the earliest origin of her line. (Brown 1990, pp. 32, 72-73, 102, 239) The 23,000-year date is the time her line left Africa via the Mideast. Sole survival of Eve's DNA requires extinction of all competing DNA. The death of all her competitors was unlikely to have been due to random causes. A single near-extinction event leaving "Eve" as the only reproductive female would be the simplest explanation for the loss of all the competing DNA.

What could wipe out all but one reproductive female 23,000 years ago? Climate cycles provide essential clues. This was not the only "genetic bottleneck" when our DNA faced extinction. Other such "bottlenecks" are known. One was 115,000 years ago and another c. 70,000 years ago. In each case, less than four thousand people survived.

But look at this: All three of these events fall near a 23,000-year cycle (115,000 = 5 x 23,000; 69,000 = 3 x 23,000; and 23,000 = 1 x 23,000). And all three designate the sudden onset of ice epochs, magnetic reversals and mass-extinctions. (Felix 1997, pp. 206-207)

Half this 23,000-year cycle of near-extinction is 11,500 years. If we count back 11,500 years, we arrive at the end of the last ice age, when global sea-level abruptly surged 375 feet. At the same time, 11,500 years ago, North America underwent a massive tectonic de-leveling event that sent a tidal wave screaming mountain-high across the Atlantic. It swamped the whole Atlantean empire and funneled a torrent of water through the Mediterranean basin, directly at Noah's world in the Middle East. Ancient people all said only a handful of us survived. Scientists insist that many animals did become extinct.

I found that I could average out the C-14 dates of extinction events over the last 100,000 years, refining them down to cycles of 11,592- and 23,184-years.

The "bottleneck" near-extinction event of the DNA-Adam and Eve c. 23,000 years ago was likewise c. 11,500 years before the well-documented mass-extinctions of ice-age mammoth and other species c. 11,500 years ago. These extinction events, we now realize, match the timing of the seven-fold Genesis Adam and Eve and the seven-fold Genesis Deluge.

The seven-fold Genesis allots exactly 11,592 years to this time-span (11,592 = 7 x 1656 cycles of time (Hebrew שָׁנָה "shaneh" = Strong's #8141; cf. #8138), if one adds up all the times of the generations from Adam to the Flood). Twice this 11,592-year cycle is 23,184 years (23,184 = 11,592 x 2). These match the results of the average C-14 extinction data.

Plato's account places the sinking of Atlantis 11,600 years ago, lining up with the seven-fold Genesis 11,592-year timing for the Deluge event. The archaeological remains of 11,600 years ago also reveal a sharp chronological demarcation between pre-cataclysm sites and post-flood sites. All but one site ends then, or begins sometime afterward, even in the Americas. (Bower 2009) The lone exception is on the highlands of Turkey exactly where an early post-flood resettlement would be expected, at a site just downhill from Ararat. (Mann 2011)

Next, consider the 92,736-year cycle (92,736 = 4 x 23,184). A 92,736-year cycle is related to changes in the shape of the earth's orbit, as it is stretched out farthest away from the sun. Typhon's close approach to the earth in late June occurs when the earth is farthest out (July 2). If we go back 92,736 years, we find yet another cataclysmic event, a sudden sea-level surge that stranded animals on hilltops where they drowned by the thousands. An ice age began afterward. (Felix 1997, p. 206) We know that the passage of a giant object like Typhon would cause similar huge ocean tidal surges and transform climate.

We have already related how the 41,000-year Milankovitch cycle of axis-tilting matches eight orbits of Typhon. Half this cycle, or 20,500 years, which Milankovitch also seems to have recognized, would be four orbits of Typhon. (Milankovitch 1972, 1930) We linked upheaval at Typhon's visit 20,500 years ago to the time in our seven-fold Genesis chronology "when mankind (terrified by Typhon) began to call upon the name of the Lord." (cf. Genesis 4:26)

We also know a great drought began 20,000 years ago, lasting thousands of years. Such a long-lasting climate change, right after a Typhon close approach, suggests that Typhon had caused a disturbance to the tilt of the earth's axis, a disturbance that would affect the planet's overall weather patterns on a semi-permanent basis. Typhon's 20,500-year cycle (4 x 5,125 years) is exactly half of the Milankovitch 41,000-

year axis-tilt cycle, a cycle also confirmed by climate data from multiple ice-core samples. (Milankovitch 1972, 1930)

Thus, in disturbing the earth's axis and changing earth's climate, Typhon is doing precisely what we would expect of such a large object and doing it at exactly the time we would expect, on the very cycles Milankovitch identified with such changes. And these changes and cycles are confirmed by both global isotope data from ice cores and by local ancient-climate data. One wonders what more we could ask of the evidence.

All of this shows that many Milankovitch cycles affecting the earth's climate are close matches to the orbital cycles of Typhon.

We have established a case for Typhon's effects on global climate, but what about its original planetary twin-brother, Typhon-2 (T-2)? Just what role has T-2 played, beyond the mega-extinction events that recur on cycles of many millions of years?

Although Typhon matches up nicely with events on the 20,500-year and 41,000-year cycles that Milankovitch found, it most certainly does not (normally) match up with the 23,184 and 92,736-year cycles. They simply do not synchronize (although we must admit that Typhon's return in our current century could come close).

That raises a question: Can T-2, arguably twice as large as Typhon (because Typhon has been damaged by colliding with the planet Uranus), be in some way connected with the basic 11,592-year or 23,184-year cycles? We need to define both orbits carefully.

First, did Typhon lose momentum in its orbit when it struck a glancing blow with the core of Uranus, some 3.2-million years ago? While it is conceivable that a sling-shot effect might have somehow occurred, and Typhon could have been accelerated, we know from Typhon's current orbit, which drives it down near the earth, that it has fallen from an orbit that had once approached no closer than the asteroid belt (before hitting T-2 some 600-million years ago). Also, had it been boosted to any great degree, we should probably have seen it far less frequently than we now do. Even if Typhon had simply bored a hole through the frozen mush surrounding Uranus, it should have slowed down.

But, there is another aspect of this event that may help explain what happened. In 2009, a new theory about Uranus was proposed by theoretical physicist Stephen Adler of the Institute for advanced Study in Princeton University (the same institution where Albert Einstein finished his career). (Choi 2009) Adler argued that the collision with Uranus was such a "colossal impact" that it lost its blanket of "dark matter" (an accumulation of sub-atomic particles thought to gradually cluster around large gravitational bodies, like the gas-giant Uranus). (Choi 2009) Adler's calculations imply that Uranus, if the impact had not taken place, would have a warmer temperature, due to the effects of dark matter. (Choi 2009) But, Uranus is even colder than the outer planet Neptune, whose distant orbit is intertwined with Pluto. (Choi 2009)

In other words, Uranus was hit so hard that it must have been violently jolted out of its inertial frame of reference, moving it away from the cloud of dark matter that had slowly accumulated around it over the course of billions of years. (Choi 2009) Moreover, there has not been sufficient time since this violent collision for dark matter to have re-gathered around Uranus. (Choi 2009) (Gribbin and Gribbin 1996)

If Adler's theory proves correct, it would support our case for a relatively recent and extremely violent impact of Uranus by a planet-sized object.

It is unlikely such a violent collision would have accelerated Typhon. It is more likely that Typhon would have lost momentum and would have fallen toward the sun, from thence to have been swung back out on its new 5,125-year orbit, where it is today.

There is further evidence for the violence of this collision. We know Pluto was also struck in the last few million years, and we argued that the Uranus event sent debris outward to disturb Pluto, explaining why it got hit around the same time. There is no requirement the two events be simultaneous, because debris from the Uranus collision could have orbited the region for some time before one of its larger chunks struck Pluto. In fact, evidence shows comets Hyakutake and Hale-Bopp may both have originated near Uranus and were kicked far away. (*Astronomy Research Notes* 2009)

Another line of evidence involves objects beyond Neptune that orbit in resonance with the orbit of Pluto. Their orbits are tilted at the same angle as Pluto's to the orbits of other planets; and their orbital eccentricity and velocity mimic Pluto ("Kuiper belt may hold fragments of Pluto," by Ron Cowen, Science News, vol. 156, Oct. 16, 1999, p. 245). It was proposed by planetary scientists at the Southwest Research Institute in Boulder Colorado, over a decade ago that some of these objects, called "Plutinos," may be debris from the violent collision that formed Pluto's moon Charon. (Choi 2009) (Gribbin and Gribbin 1996, pp. 126-127)

Ejection of debris into more energetic orbits beyond Neptune shows Pluto had been hit by an object approaching at high velocity from the Uranus side of Neptune. The more violent this impact and the more likely it came from the Uranus side; the more likely Pluto was struck by debris from Uranus' collision 3.2-million years ago. Moreover, such collision fragments disperse within a few million years. (Turner 2009) Thus, Pluto must have been hit that recently, further implicating the Uranus collision.

At least forty such "families" of objects orbit the asteroid belt. (Turner 2009) Major collision events must also have occurred there. Most asteroids come from large, differentiated objects. (Turner 2009) (Bower 2009) Their parent bodies were so big that gravitational sorting differentiated them into layers, with a core of heavier elements and, in some cases, with enough gravity to hold onto an atmosphere and have seas. (Bower 2009) (Turner 2009)

Thus, there were once planet-sized bodies in the asteroid belt that were hit by giant objects; some estimate eight Mars-sized planets were once there. (Gribbin and Gribbin 1996)

Although most people assume asteroids are simply chunks of rubble that orbit in a narrow band between Mars and Jupiter, the belt is far wider and more complex. (*Sky and Telescope* 2001) There are several bands and there are different types of asteroids in some of them. (*Sky and Telescope* 2001) (Weed 2001)

For example, what are believed to be younger, rocky silicate asteroids orbit at an average distance of c. 260-million miles from the sun (twice as far as Mars). (Weed 2001) The more primitive type of asteroids, believed to date to the earliest days of our solar system, are the "carbonaceous chondrites," which tend to orbit 300-million miles out (as far from the average orbit of the younger asteroids as we are from Mars). (Weed 2001)

While the sun and Jupiter's gravity might have sorted asteroids by their size and mass, it is harder to explain why objects might have been sorted by their chemical make-up. (*Sky and Telescope* 2001) This suggests that orbital sorting relates to where these asteroids fragmented. This idea, combined with the different ages of the two types, strongly argues for two major debris-producing events in the asteroid belt.

The earliest event, before 4-billion years ago, produced the carbonaceous objects with round chondrules of embedded material. However, there are a number of puzzles. All these objects originated in the asteroid belt. But, each carbonaceous chondrite contains two kinds of material, some of which has been previously heated near 2,500° F., while other material has supposedly never been heated at all. (*Sky and Telescope* 2001)

This seeming contradiction is only part of the mystery. These chondrules have been heated to those high temperatures for extremely brief times: between a few minutes to a few hours, or at most, days. (Sky and Telescope 2001) This is known because various volatile elements in them did not have time to evaporate, even in the extreme heat. (Sky and Telescope 2001)

There's more: Some chondrules absorbed short-lived radioactive material and they all solidified in a strong magnetic field. (*Sky and Telescope* 2001)

Because it is not accepted by most scientists that a large planet was in the asteroid belt, the theory of the controversial astronomer: Dr. Tom Van Flandern that a Saturn-sized planet there had blown itself apart is never considered. Yet, it is the obvious explanation for all these details. If a large planet were to be spun apart, there would be brief heating of surface rocks and an exposure to radioactivity from inside the planet. Mixtures of material from unheated and briefly-heated debris would combine during the minutes and hours following the event. (van Flandern 1993)

This material tends to be found *c.* 300-million miles from the sun, in the outer part of the asteroid belt. This is closer to Jupiter, which is thought to have been somewhat nearer the sun at that time, in a place where it may have interacted with the Saturn-sized gas giant then in what became the asteroid belt. There was originally a thousand times more material in the belt, enough to form a rocky planet several times the mass of the earth, (*Sky and Telescope* 2001) or up to 1,200 times more. (Gribbin and Gribbin 1996, p. 95)

The other group of asteroids, those that formed later, reside closer to the sun, about 260-million miles out. These rocky silicates are made of materials that resemble the inner part of a planet. In that case, they might have resulted from the collision of the two halves of the original planet's fractured core.

Some explain the melted "chondrules" in the oldest asteroids as flash-heating by the sun and flash-freezing by a solar wind that was thrusting particles away. (Beatty and MacRobert 2001) But, since some "chondrules" were heated for but a few minutes, they had to be hurled away from the sun at extremely high velocity, beyond 25,000 miles per hour, the escape velocity of solar gravity. They would have been ejected far beyond the asteroid belt. Moreover, while some particles were melted for minutes, others were heated for hours and a few for a day or more, yet these disparate particles always arrived at exactly the same time and place, ending up mixed inside the same asteroids with totally unheated material.

By contrast, the planet-expulsion hypothesis matches the data perfectly. Saturn is 75,000 miles in diameter. A planet of similar size could have briefly existed in the asteroid belt, based upon the amount of original mass scientists calculate to have been in the belt. This second Saturn would have had a radius of roughly 37,500 miles. Material ejected from it during the break-up could have been moving at less than 25,000 miles an hour yet be heated for a matter of minutes or a few hours during the explosive final phase of the destruction. It would depend upon how deep in the planet the material had been located. Multiple melt-times and temperatures would be expected, as well as some radioactivity and magnetic field exposure, exactly as found in the samples. (Beatty and MacRobert 2001)

Afterwards, the melted droplets would have entered a surrounding debris field of spun-out cold surface material. The mixing of the two would produce the observed matrix of cold and melted carbonaceous chondrites. The presence of unheated carbon and water implies exterior and atmospheric regions of a cold original planet. The original orbit of that planet would have been in the asteroid belt, *c.* 300-million miles from the sun.

On the other hand, the presence of two distinct bands of asteroids with two very distinctive compositions, the inner one being younger, suggests two events in which large clouds of debris were generated.

And, in fact, two massive debris-producing events are known to have occurred in the history of the asteroid belt. The first was the early solar system bombardment that peppered the planets some four billion years ago. (Gribbin and Gribbin 1996, pp. 61-70) We have identified that with the original break-up of the Saturn-sized gas giant.

The second event was roughly 600-million years ago, assuming that as the time of the more recent great bombardment, when life began to flourish. (Gribbin and Gribbin 1996, pp. 61-89) (Felix 1997, p. 38) We would identify this second cataclysm with the meteorite bombardment that occurred when the two halves of the exploded core of the Saturn-sized planet eventually collided, during one of their orbital returns to the asteroid belt.

But, only the second of these was an actual collision. So what else do we know about this second event, some 600-million years ago?

From the end of the early bombardment 3.8-billion years ago, until the sudden increase in impacts 600-million years ago, there had been a long, slow gradual decline in cratering. (Gribbin and Gribbin 1996, pp. 88-89, note 1) The Moon had remained fairly unmolested during the intervening 3.2-billion years. Then, 600-million years ago, the Moon began to experience intense new bombardment, with nearly 6,000 "young" craters since that time, a major hit every million years or so. (Gribbin and Gribbin 1996, pp. 64-65)

The Moon was, of course, not the only target. Venus was peppered with such an intense new bombardment 600-million years ago that the entire surface of that planet was blasted clean in the cataclysmic melting resulting from those events. (Gribbin and Gribbin 1996, pp. 73-74) As with the earth, the impacts were so horrific that Venus erupted in volcanic lava that began to flood over the whole planet, wiping out all its earlier impact history. (Gribbin and Gribbin 1996, pp. 73-74)

The earth also began to be heavily pounded 600-million years ago. (Gribbin and Gribbin 1996, pp. 73-74) This time-frame coincides exactly with the period when fossils first appear in profusion in the geologic layers. (Gribbin and Gribbin 1996, pp. 64, 73) The implication that the bombardment itself has not only caused the fossil deposits through cataclysmic burial, but that the side-effects of the bombardment (mutagenic gases, ozone layer destruction, selective extinctions) may well have caused the development of new life-forms. Immanuel Velikovsky first suggested in 1950 that cosmic cataclysms triggered evolutionary changes. (Velikovsky 1950, 1967, p. 388) In this light we may recall that another name for Typhon was "the outstretched arm" of God, the 'hand' by which God judges the earth.[1]

It had been thought that Mars had not had a 600-million year ago bombardment. (Gribbin and Gribbin 1996, p. 74) But, more recent studies suggest that a secondary series of impacts has in fact occurred; and it may be that a super-massive impact with a small-planet-sized object has wiped the original impacted surface clean. (Kiefer 2008)

It has been said that carbon-dioxide levels on the earth at that time hit a peak of 20 times the current level, although sea-levels were then relatively low, possibly even lower than today. (Felix 1997, pp. 49-50)

The primary source of atmospheric carbon dioxide at that time, when there was hardly any animal or plant life on the land, was apparently volcanic activity. Volcanoes vent enormous amounts of carbon-dioxide when they erupt. We now know of thousands (or millions) of volcanoes on the sea-floor that vent carbon dioxide.

After the collision of the two core remnants 600-million years ago, the earth was brutally bombarded with debris, as was the rest of the inner solar system. It is not at all unlikely that the extreme level

[1] Isaiah 1:2, 25-31, 2:2, 19-21, 5:25, 30, 9:17-21, *etc.*

of carbon-dioxide resulted from massive volcanism triggered by the violent pounding of the earth after the collision. (Felix 1997, 31, 36)

Moreover, 600-million years ago would also have been when Typhon and its twin, T-2, made their first close approaches to the earth. Prior to their collision, both Typhon and T-2 had in theory gotten no closer to the earth than the asteroid belt, the site of the c. 4-billion-year ago break-up of the Saturn-sized gas-giant. After colliding with each other 600-million years ago, both Typhon and T-2 began to make close approaches to the earth, with T-2 swinging past the earth in the same direction as the earth's orbit.

How do we know this? Recall that T-2 made a super-close flyby 65-million years ago, erupting massive lava out of the rift in the south Indian Ocean. It lifted Pakistan, India, Bangladesh, Indo-China and much of Indonesia, and dragged them uphill at super-sonic velocity (tsunamis of 400 mph) toward the southern flank of Asia. (Felix 1997, p. 30)

The stentorian sound of this event could, by itself, have killed dinosaurs. Tidal waves of scalding hot, acidic salt water spread out in a vast fan of death ahead of the Indian subcontinent as it plowed northward through the Indian Ocean. (Felix 1997, p. 31) (Gribbin and Gribbin 1996, p 34) The boiling tsunamis, bubbling with sulfuric acid should have created a huge arc of deserts as they struck land. And deserts still form an arc around India's path, from East Africa and Arabia to Persia and the Takla Maken Desert and the Gobi Desert in China, on over to the Australian outback, precisely where we would have expected from such monstrous tidal waves of sulfuric-acid and sea-water. (Felix 1997, p. 31) (Gribbin and Gribbin 1996, p 34)

Beyond what became the Bay of Bengal, T-2 continued passing over the Arctic, Canada, and Iowa, (Gribbin and Gribbin 1996, p. 32) (Lewis 1996, p. 143, though this author disputes the Iowa date) casting its final blow at the Yucatan. This direction of motion is determined by India being dragged uphill in a brief hour or less, so quickly neither the Indus nor Ganges Rivers left an alluvial deposit on the floor of the Indian Ocean until India collided with Asia. The huge size of the post-collision alluvial deposits attests to the vast time India has resided in its current location.

It would seem that only a nearby massive cosmic body traveling northward over the Indian Ocean would be able to explain this situation. India was erupting with waves of lava at the time, 65-million years ago. This fountain of lava erupted out of the south Indian Ocean's rift, an enormous lava flow plainly visible on the sea-floor even today. The Indian sub-continent erupted with lava at the same time as the major impacts from Siberia to the Yucatan, and from the same cosmic cause. (Gribbin and Gribbin 1996, p. 32-33)

This passage of T-2 had moved so much material from one part of the earth to another that the earth's rotation may have become unstable. If so, the axis of rotation had to re-set. We know there is an alternative South Pole in the south Atlantic and that it has its twin alternative North Pole in the north Pacific. These may have been the poles prior to the extinction. Afterward, they could have re-set to something like the location of our current poles (although the sites have alternated many times since).

An alternative pole prior to the extinction made the path of T-2, not northward, as it would be today, but what would then have been north-westerly. After T-2 came by, the north Pacific pole had shifted up toward the current Arctic pole site.

Could the old North Pole have been the pole in the south Atlantic? It seems less likely that the old North Pole was as far south as the south Atlantic, because a re-set of the poles would probably not have been that extreme but would have stopped as soon as it found any stable new position.

If we can accept that T-2 was then moving westerly as it passed by the earth, we next need to know if it was passing the sunward or night-time side of the earth. Scientists believe North America was then in noonday sunlight. (Gribbin and Gribbin 1996, p. 33) But, this assumes an ordinary comet had broken up near the sun and that its pieces hit the Yucatan first. (Gribbin and Gribbin 1996, p. 33) However, it is extremely unlikely that comets just happen to clobber the earth on a 65-million-year cycle. Moreover, we have presented evidence that the sequence was reversed: The impact in the Yucatan was probably the last, not the first, that day.

If so, then the direction was opposite to that which these scientists have assumed. Also, the impacts were probably not caused by an ordinary, randomly-occurring comet break-up, but rather, by objects from T-2's large debris cloud, leftover from its collision with Typhon, and having a predictable 65-million extreme close-approach cycle.

When it first affected the Indian subcontinent, T-2 had to have been approaching the earth on its far side from North America. So, it would have been on the dark side when it arrived, if North America were then in sunlight. In that case, T-2 was moving retrograde, opposite to the direction of the earth as it moves around the sun.

On the other hand, given that all of the assumptions about direction that have been used by scientists to describe this event seem to be opposite to what happened, we may be on safer ground to infer that North America was in darkness. In that case, T-2 was not moving retrograde, but was actually moving pro-grade, that is, in the same direction as the motion of the earth. A pro-grade westward motion corresponds to an object passing by the earth on the sunward side. That would mean that India was moved in daylight.

Accordingly, it appears T-2 approached the earth on the sunward side, arriving in an epoch when the poles were about 45 degrees off from their current locations. T-2 passed over India, which was then in the south Indian Ocean. The gravity of T-2 caused the mid-ocean rift under India to explode with lava, which gushed up onto the western flank of the subcontinent, exactly at the longitude where the mid-ocean rift can be seen today, albeit far to the south of India's current position. (*National Geographic Society* 1967)

The erupting lava greased the underside of the Indian subcontinent, allowing it to slide freely down off the rift and move over the sea-floor, accelerating rapidly as it was lifted by the grasp of T-2, moving overhead, in what was then a north-western direction.

To be certain of this orientation, I simulated the event on a large globe. I already knew Typhon-2 must have been moving in the ecliptic to have an orbit that intersected both the asteroid belt and the earth. And so, it could only have been pacing with the earth on its sunward side, at a time when the North Pole was in the north Pacific Ocean, south of Alaska. I also found that the earth at that time would have been tilted toward the sun, making it near the Summer Solstice in the Northern Hemisphere.

I was surprised to find that Typhon and Typhon-2 both seem to come by the earth at the same time of year. But upon reflection, it makes sense. The tilting and orbital alignments with the Summer Solstice would not be as consistent as they now are if the two objects were currently competing to tug the earth in different directions.

On the other hand, by the process of elimination, when the dinosaurs died, Typhon had been moving in the other direction during its passages, that is, retrograde.

But after colliding with Uranus 3.2-million years ago, Typhon would have been diverted into a new orbit. Moving much like a comet, Typhon would have been far enough out to 'pick' which side of the sun to swing around after it hit Uranus. Depending which side of the sun it ended up going around, Typhon could have changed its direction of motion relative to the direction of T-2.

That means both Typhon and T-2 would now be in pro-grade motion. If that is indeed the situation, then how fast could they be moving when they go by the earth?

When I saw Comet Hyakutake as it came by in late March of 1996, it was in a 4,000-year long orbit, similar to Typhon's 5,125-year cycle. At midnight on the night it went by the earth, I got to see Hyakutake at its closest approach. I watched it for several minutes and could detect no forward motion, although it was traveling at an extremely high velocity. The next night, it was still there, albeit smaller and further along.

It seems odd, but earth is so large that an object passing us would take time going by. The earth is approximately 8,000 miles in diameter. For example, even if moving at a relative speed of 4,000 mph, it takes an object two hours to pass us, moving the same direction. But an earth-sized object (one that is also 8,000 miles wide) would take another two hours for all of it to move by: four hours in all. And T-2 might be twice as big as that.

T-2 is so massive that it could have had a gravitational force strong enough to move India even at greater than lunar distance. The fact that the moon is still in a fairly circular orbit attests to T-2's probably only rarely, if ever, coming as close as the moon.

If it were farther than the moon, the effects of a T-2 passage could still be felt for well over eight hours, even if it were moving 4,000 mph faster than the earth.

But the relative velocity of the passage might have been different, and that would produce a somewhat longer or shorter passage time, but still measured in several hours.

Now, we can begin to ask if T-2 is responsible for the 23,184-year cycle. Five of Typhon's 5,125-year orbits are about 153 years shy of the 25,780-year precession of the equinoxes. (Houghton, Mifflin, Harcourt 2016, p. 1030, my calculation is based upon a precessional rate of 50.27 arc-seconds/year) But the 23,184-year cycle is 2,600 years shy of this precession cycle.

However, rate of precession is accelerated by the drift of the earth's orbital shape, a gradual rotation of the long-axis of our orbital ellipse. (Felix 1997, p. 178) The net effect is to reduce the total elapsed time of the earth's 25,780-year precessional circuit to only about 23,000 years, which seems to correspond to the 23,184-year cycle. (Felix 1997, p. 178)

However, as we said before, the 23,184-year cycle rarely coincides with Typhon's passages. That would suggest T-2 is the mechanism keeping the 23,184-year cycle going. By comparison, comet Hale-Bopp had a c. 20,000-year orbit, but it was thought to have come from the farthest reaches of the solar system. It is doubtful that T-2 has an orbit that far out, especially after colliding with Typhon 600-million years ago.

Is T-2's orbit some smaller length, divisible into 23,184? We have already found a destruction cycle of 11,592 years. If T-2 has an 11,592-year orbit, it must be in the plane of the ecliptic, or close to it, because T-2's orbit 'began' in the asteroid belt, yet it makes regular close approaches to the earth. This creates data points that have to fit into a solar-centered orbit. The only way to make them fit, as with Typhon as well, is for the orbits to be almost exactly in the same plane as the earth's orbit, that is, in the ecliptic.

Like Typhon, T-2 should enter the Kuiper belt each orbit. Scientists say a second earth-scale object may have helped sculpt the Kuiper belt. (Trujillo 2003) So, at least two earth-scale planets, like Typhon and T-2, in comet-like orbits, were moving in the ecliptic, shaping the Kuiper belt. Why imagine there were two additional planets when we already have evidence for Typhon and T-2?

Now we have a much better understanding of T-2's orbit. It must be within our orbit when it makes its closest approach to the sun. Unlike Typhon itself, however, it does not come close to the earth on every passage. If it did, we should have experienced the horrors of the dinosaur extinction event every 11,592 years.

One of the most terrifying things that happened to the world of the dinosaurs was the global firestorm, whose flames raced around the world at jet-speed, accompanied by sulfuric acid tornadoes. (Gribbin and Gribbin 1996, p. 34-38) The dinosaurs were suffocated, burned to a crisp, and most remains were devoured by acid. (Gribbin and Gribbin 1996, p. 34-38) Tornadoes and tidal waves swept up the bulk of their ashes from the land and rained them out, flushing them into the sea.

The skies darkened for years. Besides things like insects, only the mammals, who normally came out in darkness to feast on dinosaur eggs, were able to escape, along with the smallest of the flying dinosaurs, ancestors of modern birds. Mammals probably spent their days asleep in the recesses of caves, where they hoped to be safe from the clutches of daylight-feeding dinosaurs. Large dinosaurs could not hide in small caves.

It happens that a chicken egg (that is, the egg of a descendent of a small flying dinosaur that could take refuge in caves) contains the perfect balance of protein for humans. So, eggs, which are now probably the most inexpensive and easiest source of protein, are a relic of why mammals like us were able to survive the age of the dinosaurs. During the dark years after the extinction, such bird eggs became a primary source of protein, without which we would not be here today. This suggests that our genetically-adapted love of eggs is a side-effect of Typhon-2's destruction of the dinosaurs.

T-2's effect on our modern world may be far more dramatic than we would have imagined. When I examined the global sea-floor maps, looking for clues to T-2's effect on India, I kept puzzling about the locations of the various continents. Even after allowing for mid-ocean rifts supposedly spreading continents apart, I could not explain why North America was over-running the Pacific rift. Also, Africa appeared to be too far east and north, even twisted a bit. And, Asia was way too far to the west of the Pacific rift, entirely disconnected with its spreading region.

Something besides spreading rifts clearly seems to have moved the continents. When I looked at Africa, I could see it had been twisted in the direction of India. That would make perfect sense. If T-2 had dragged India northward, Africa could have been carried north as well. Africa had also been thrust into the Asian continent, but being already near it, the impact was not as severe as it was with India.

Then, I looked at North America, the last continent T-2 had passed over as it flew by the earth. It is farther west than it should be, but I noticed that Greenland looked as if it had been torn out of the Arctic and wrestled into the Atlantic. These motions match the general direction of T-2's passage 65-million years ago. Keep in mind that, at the time of its passage, K-2's motion would have been relative to where the poles were then located.

All continents are "greasy" with lava on their undersides. They slide freely over the sea-floor, as if it were ice. But the sea-floor is flexible, like a half-frozen water-bed. The continents cause it to sag. When the Americas were shoved west on another occasion by T-2, the heavy continents rode up over the crust of the Pacific sea-floor, creating a subduction zone on the west coast of the Americas. One scientist suggested Antarctica was pushed to the South Pole by a gigantic comet or cosmic object when the ice ages began 3.2-million years ago. (Gribbin and Gribbin 1996, p. 150) And Typhon began its close passages with the earth at that very time, after hitting Uranus and falling into a new *c.* 5,125-year orbit.

Asia has been moved as well. The volcanism and earthquakes along the east side of Asia occur far from the Pacific "mid-ocean" rift, which is nowhere near the middle of that ocean. Asia clearly has been moved a great distance westward from its mid-ocean rift. In fact, it is on the far side of the ocean from where the rift is, which now dives under North America. The lava ridges west of the Pacific rift under North America end in the midst of the Pacific Ocean, thousands of miles from the eastern edge of Asia.

Obviously, the rift did not push the huge mass of the Asian continent to its current location. There is no sea-floor spreading between East Asia and the middle of the Pacific Ocean. But the immense gravitational attraction by T-2 is more than capable of exerting the force necessary to slide the great continent of Asia far across the Pacific Ocean.

Continents all over the world have probably been shoved around, ruptured and broken apart by T-2. Prior to hitting Uranus 3.2-million years ago, Typhon may or may not have added to T-2's remaking of the earth. In any case, our coastlines and mountain ranges, our orientation to the earth's rotation, and thus, our climate, have all been determined by how T-2 has broken up and moved our lands around.

We can find even more evidence for our thesis by measuring and back-tracking the global patterns of the undersea faults and rifts. All three rift systems--the Atlantic, the Pacific and the Indian Ocean rifts--are cut across by massive sideways faults. The rifts look as if they were sea-serpents sliced up by a sushi chef. These slicing faults occur with regular frequency along most of the length of these massive "mid-ocean" rift systems.

Such a regular global pattern is not caused by random local quake activity. Virtually every one of these rift-hacking faults is far longer and deeper than almost any fault we see on land. The whole sea-floor is carved up with enormous gashes that could swallow a little island like Manhattan without burping.

But, there is a vivid and unmistakable pattern to these massive gashes. They radiate outward from two central focal points thousands of miles away from the main portion of the rifts. The focal point of the faults that crisscross the Atlantic and the Indian Ocean rifts is roughly between Africa and the island of Madagascar, a giant chunk of land that had once been attached further up the coast of Africa. (Svitil 2000)

Many millions of years ago Africa was located differently than it now is, a bit to the southwest, and rotated slightly toward the northeast. (Svitil 2000) At that time, before the dinosaur extinction, the site of what is now Lake Victoria was centered on the focal-point of all those giant fault-lines that carved up the Atlantic and Indian Ocean rifts.

Lake Victoria is not just the main source of the Nile. Not far to its east, one can see the snows of Mount Kilimanjaro, over the border in Kenya. Kilimanjaro is nearly 20,000 feet high, the highest point in Africa. From its peak, I suppose, some ancient flying creature might long ago have witnessed the break-off of Madagascar from the African coastline. But Kilimanjaro is part of an enormous ring of mountains that cradle Lake Victoria, along with adjacent water basins of a peculiar crescent shape, with which Victoria is embraced on each side. The feature appears to be that of a colossal volcanic caldera, far greater than Yellowstone.

Out of the mouth of this Victoria caldera we can trace two rift valleys, each of which looks just like the ones that supposedly cause ocean-floor spreading, right out in the open, on land.

One of these goes north to form the Red Sea, the Gulf of Aqaba, the Dead Sea (the lowest point on the continents), the Jordan River Valley, and the Sea of Galilee. The western ridge complex of this "Great Rift Valley" includes the Mount of Olives overlooking Jerusalem and, arguably, the Temple Mount itself.

The other rift is even longer, for it becomes the Indian Ocean rift and then, as it snakes around the southern tip of Africa, becomes the Atlantic Ocean rift. And all the while, as it meanders along the seafloor, the rift is sliced open by faults that point back to where the gaping Victoria caldera once coughed up trinkets like Kilimanjaro.

It should be quite obvious that the Victoria mega-volcano once lay at the heart of a vast continent of out-poured lava that we now call Africa. In addition to Africa, the molten gusher should have generated the lost half of Africa once attached to its east, including Madagascar, India, Pakistan, Antarctica, Australia, and New Zealand. (Svitil 2000)

The fault lines of Atlantic Ocean seafloor rifts that are half a planet away point back at Victoria, like bony fingers accusing their perpetrator.

These earth-straddling scars imply events of such globe-shattering magnitude we are compelled to seek a cosmic cause.

Yet, there is an older gouge in the planet. The Pacific seafloor rift arcs around the Pacific basin, beginning in the Indian Ocean, and slithers up under California and Nevada, northward between the Cascades and the Rockies into Alaska, where a feature like Victoria can be seen, not far from Mt. McKinley, the highest point in North America.

But that may not be the end of it. Not far from there it seems to reappear as a deep trench, the deepest on earth, diving almost into the molten mantle by the time it reaches the Philippines, which are currently located near the focal-point of all of the ancient Pacific Ocean rift fault-lines, including those under America and in the Indian Ocean.

Here we may note that this location is exactly 90 degrees from the center of the Victoria caldera. Not only that, but it appears to be as far north of today's equator as Victoria was then south of what is now the equator.

The whole picture seems perfectly symmetrical. Makes you wonder, does it not, if there might be other similar patterns? Your quest would be richly rewarded.

For example, the two focal-point centers of the mid-ocean ridges were about 22.5 degrees from the equator. That is, Victoria was near the Tropic of Capricorn and the Philippine center is near the Tropic of Cancer. These lines indicate the degree of tilt of the earth's axis today. Remember that we found Typhon's 5125.36-year cycle, times 8, echoes the Milankovitch cycle for the changing degree of tilt of the axis: 41,000 years. This would be consistent with a cosmic body causing the tilt of the axis. It also links the passages of Typhon and the times of the solstices (which are determined by the axis tilt).

Another example: The Victoria and the Pacific focus-points appear to have been located on a straight line (that is, a great circle route) equidistant between the alternative poles, the one in the northern Pacific and the other in the south Atlantic. These four points would define a line of longitude whenever the earth switched to the alternate poles.

Or consider this: The alternate poles produce an alternate equator, and the Dead Sea, the lowest point on the surface of the earth, was not only on it, it is now due north of Lake Victoria. That alternate equator bisects the Victoria/Philippine alternate longitude line at a long ridge (drag-mark?) under the Indian Ocean that points due north at India.

These mega-encounters with gigantic cosmic bodies do more than re-arrange the continents and tilt the axis. In the great extinction 65-million years ago, at least a quarter of the total bio-mass of the whole planet was incinerated to soot, but even more could have been totally consumed to pure carbon dioxide, whose concentration suddenly tripled (Gribbin, pp. 35-37). The location of the soot in the clay layers revealed that the fires ignited even as the debris from the impacting comets began falling back to the earth, that is, almost instantly, while these impacts were happening. (Svitil 2000, p. 35)

This sounds like our modern fireball activity. Yet, some meteor specialists seem reluctant to admit that fireballs set fires. (*cf.* http://amsMeteors.org) Is there evidence that fireballs and flaming meteorites can ignite fires?

Absolutely... Dr. John S. Lewis, Professor Emeritus of Planetary Sciences at the Lunar and Planetary Laboratory and Co-director of the NASA/University of Arizona Space Engineering Research Center, a scientist who has specialized in meteorite and comet events, compiled in the late 1990's a long list of reported damages caused by impacts during the past 2,000 years. (Lewis 1996, pp. 158-182) The list was by no means complete. (Lewis 1996, pp. 223-225) He found at least 23 cases where these objects set fire to things on the earth, including entire cities. (Lewis 1996, pp. 178-179, 225)[2]

Dr. John S. Lewis

That "cosmic fire" is something peculiar to Typhon and T-2 can be seen by comparing the effects created by Mercury's impact on Mars. There is no sign of burning over the surfaces of either Mercury or Mars. While volcanic activity on Mars may have occurred then, and although there is ash around the rim of Mars' Olympus Mons volcano, the rest of the surface of Mars seems unaffected by burning. This may be largely due to the absence of oxygen, of course, but oxygen could have been freed up by the super-heating of rocks or reconstituting sub-surface water during the collision.

Scientists largely accept that the moon was created by a Mars-sized planet striking the earth. They also realize that Mercury used to orbit much farther out from the sun before it collided with another planet. (Ferris 1997) Timothy Ferris, a *professor emeritus* at the University of California and Prolific Public Broadcasting System (PBS) filmmaker, has suggested that Mercury was not only hit by another planet, but that the earth may itself have had three other huge impacts caused by "hefty" objects that were almost planet-sized. (Ferris 1997) Mars has also been shown to have had a massive impact on what is now its North Pole; this excavated the northern half of the planet and dumped much of the rubble on the southern half, which is uniformly higher. (Goldberg, *et al.* 2014) Uranus, as we know, was hit by an earth-sized planet and knocked over onto its side.

[2]*Ed. Note:* Dr. Lewis is also Chief Scientist for Deep Space Industries, a private company researching ways to mine minerals in outer space.

Today no one doubts planets hit each other, or that extremely violent events have happened repeatedly during solar system history as planets wander out of safe orbits.

It is now predicted that Mercury itself is going to one day stray out of its current unstable orbit and have a close passage with Venus that will send it out to threaten the earth again. Mars is also going to leave its orbit. (Gribbin and Gribbin 1996, p. 143 note 3)

It is ironic that the late Immanuel Velikovsky was vilified half a century earlier for daring to argue for such events, presenting hundreds of pages of documentation. His views may not have been flawless, but the ideas he put forward regarding Mars and Mercury (in a book posthumously published on-line) have repeatedly been supported by later discoveries. Yet, as late as 1994, some scientists were still convinced that not even small comets could never hit Jupiter or the sun. (Carlisle 1995, p. 88) It is surprising how dogmatic top experts in the field had been when asserting things that no scientist today believes.

Mercury has had at least two big collisions. The first one stripped the planet of its outer rocky crust. (Carlisle 1995, p. 88) I believe this occurred when Mercury hit the earth and formed our moon. Sometime later, perhaps much later, Mercury had another big impact, one that fractured the entire far side of the planet and left a huge circular zone of debris at the impact site, the Caloris Basin. (Fecht 2011) It is this site that I identify with the huge impact on what is now the North Pole of Mars. At the time Mars and Mercury were both much nearer the earth, and this event sent the two of them into their current orbits.

However, in the millennia that preceded this collision, Mars and Mercury would take turns passing near the earth, occasionally raising huge tides and causing earthquakes, world-wrecking storms, and rains of dust and debris (Velikovsky, *Worlds in Collision*, where I would substitute Mercury for the supposed Venus events).

These destructive encounters never produced the kinds of extinction-level events or global firestorms that characterize Typhon and T-2's extreme passages. So, it would appear that having a large, planet-sized mass is not the only thing responsible for the colossal level of havoc caused by Typhon and T-2. Another factor is involved, and that would seem to be the peculiar nature of the debris field that is associated with these objects. What is in this debris field?

The debris we can study most easily would be that associated with the most recent Typhon passage just before 3100 B.C. It happens that two prominent British astronomers, Victor Clube and Bill Napier, have already steadied the supposed break-up of a "gigantic" comet related to this. (Clube and Napier 1982, pp. 131-285) They carefully analyzed dozens of comets and asteroids that seem to have derived from a cataclysmic event 5,000 years ago. (Clube and Napier 1982, p. 149)

Among their observations were that short-period comets (with orbits under a few thousand years) are 150-times too abundant, unless something recently injected a huge number of new comets into the solar system. (Clube and Napier 1982, pp. 131-132) They cite a dozen comets with orbits of several hundred years each that can be traced back specifically to "a single gigantic object" passing near the planets 10,000 and 20,000 years ago. (Clube and Napier 1982, p. 133)

We need not wonder what happened. Typhon came by 5,000, 10,000 and 20,000 years ago, as we have seen. We also found hard evidence for its fiery visit 15,000 years ago. It could have dumped out these twelve comets on any one of those visits.

Clube and Napier also argued that the asteroid belt should be as flat as "the rings of Saturn," not look puffed up like "a doughnut." (Clube and Napier 1982, p. 67) One possible explanation for this they considered (but dismissed) was a disturbance of the belt by "a very large number of massive bodies." (Clube and Napier 1982, p. 67)

Alternatively, of course, repeated passages of two Typhon-sized objects through the asteroid belt could have puffed it up, just as with the Kuiper Belt. Note also that there is no need to invent two new objects since we already have evidence for Typhon and T-2.

Next, Clube and Napier said earth-crossing Apollo asteroids are ten times more numerous than can be explained by ejections out of the asteroid belt. (Clube and Napier 1982, p. 76)

Either they come from outside the planetary system or there must be something from outside disturbing the asteroids. Once again, Typhon and T-2 orbit like comets and pass through the asteroid belt; they explain the excess Apollo earth-crossing objects.

Another observation by Clube and Napier was that there are twice as many earth-crossing asteroids as would be expected by the number of impact craters in the geologic record. In other words, we currently have twice as many asteroids threatening the earth as the record of the last several hundred million years would lead us to expect (Clube and Napier 1982, p. 84). Knowing Typhon recently hit Uranus and is bringing a cloud of impact debris with it provides a simple explanation for the sudden influx of Apollo asteroids. We tied Typhon's increased impacts to the onset of ice ages over the last 3.2-million years.

The moon has no atmosphere to limit impacts, nor weather to wear away craters. Clube and Napier said there are five times as many Apollo objects as ought to be orbiting near us, based upon the number of lunar craters. (Clube and Napier 1982, p. 84) Too many Apollos have recently begun crossing our orbit. Typhon has apparently brought them in and deposited them in our orbital region.

Yet, another puzzle identified by Clube and Napier is that the vast majority of the meteorites that have been recovered and studied by scientists appear to have come from bigger objects, large asteroids or even small planets. (Clube and Napier 1982, p. 86)

Of course, they did not consider objects as large as Typhon and T-2.

They pointed out that there has been a long, steady geometric decline in the rate of cratering since c. 4-billion years ago. (Clube and Napier 1982, p. 55) It exactly parallels what one would expect from the catastrophic break-up of a Saturn-sized planet at that time and the geometric decline of impacts as the resulting debris field steadily dissipated.

A second "layer" of cratering and gradual decline began c. 600-million years ago, but the rate of decline has been flatter. (Clube and Napier 1982, p. 55) Again, the peak corresponds to Typhon and T-2 colliding at that time, but the flattening out of the rate of cratering corresponds to the additional number of asteroids and comets being transferred into earth orbit since then.

There is still more evidence. They had noticed in the early 1980's that there was already an "overabundance of fireballs" hitting the earth. (Clube and Napier 1982, p. 137) We have seen that not only are there more fireballs than expected, but that the number is increasing year by year (a fact Clube and Napier did not apparently realize). Worse still, the numbers of new fireballs are growing exponentially, indicating that the return of Typhon is fairly immanent on cosmic scales, perhaps less than a decade away.

Also, the supply of dust between the planets is a breathtaking 10-to-100 times too great if it comes from the current number of observed comets. (Clube and Napier 1982, p. 137) Clearly, there must be another source for all this dust. Clube and Napier estimated that a gigantic comet had broken up about 10,000 years ago, seeding the inner solar system with the dust needed for rain and the agricultural revolution. (Gribbin and Gribbin 1996, pp. 151-159) Knowing that Typhon hit Uranus and scooped up a huge quantity of dust resolves this issue.

Other data Clube and Napier investigated led them to infer that either there has been a recent large increase in the comet population, or "there has been a (single) giant comet (near earth) in the recent past" (Clube and Napier 1982, p. 142)

Curiously, like us, they tied the Tunguska Siberian comet impact to the fireballs (impacted June 30, 1908), based upon an analysis of energy released and frequency. (Clube and Napier 1982, p. 142) That is, they found statistical evidence that Tunguska, the largest recent Typhon debris impact, has the same origin as the fireballs.

Since April of 1974, there have been four seismic detectors on the moon, left by the Apollo astronauts, counting the number of objects striking the lunar surface. (Clube and Napier 1982, p. 148) The data from these detectors provide unambiguous evidence about impact rates during the year. (Clube and Napier 1982, p. 148) The data show that impacts on the moon peak in late June during the Taurid meteor stream. (Clube and Napier 1982, p. 149) That is the same time as the 1908 Tunguska impact, and it is the same time as the recent rapid increase in earthly fireball reports.

Moreover, that lunar late June peak is three times higher than the rest of the year. (Clube and Napier 1982, p. 149) They calculated that the Taurid meteor stream originated about 5,000 years ago with "an exceptionally large progenitor" (the break-up of a gigantic comet). (Clube and Napier 1982, p. 149)

The key detail is that this 5,000-year old Taurid stream comes by the earth in late June. There were no other peaks except the other Taurid crossing in late October and early November, which was double the background level.

Clube and Napier then compiled all the parallel evidence for the Taurid meteors, for comet Encke, for the Tunguska 1908 object, and for the Apollo asteroid Hephaestos. They concluded that these all had derived from a single gigantic source object 5,000 years ago. (Clube and Napier 1982, pp. 151-153)

This is essentially the Typhon "smoking gun" evidence. It means there is empirical evidence that there was a huge object with a large reservoir of comets and fireball debris that came into our solar system about 5,000 years ago, that is, near 3100 B.C., the era of Typhon's most recent visit, the origin of ancient civilizations, the time of Menes and his Minoan empire, the start of the Mayan calendar, *etc.*

Furthermore, Clube and Napier found evidence that this object had come by as far back as 20,000 years ago. (Clube and Napier 1982, p. 133) Obviously, they were describing something that had come by repeatedly, dumping comets and other debris, including cosmic dust, into the solar system on each pass. And the timing shows that it has a 5,000-year orbit.

This is a precise description of Typhon. And as it turns out, they finished their discussion with a study of references to such a giant destructive comet in ancient literature. Over and over again, one name kept recurring in their investigation of ancient texts: Typhon. (Clube and Napier 1982, pp. 185, 192-211, 220, 257, 262, 272; *cf.* 172*ff,* 255)

We also have new DNA evidence for the near-extermination of mankind at the time of Typhon's last passage *c.* 3100 B.C. Genetic researchers recently took DNA from over 2,000 volunteers and examined a segment of their genetic code to determine the rate of mutations and how far back they began. (Tennessen, *et al.* 2012) Their research found a sudden explosion in human genetic diversity occurred about 5,115 years ago. (Tennessen, *et al.* 2012) This would correspond to about 3105 B.C.

In order for this sudden spurt in genetic diversity to have its origin at that time, the human race had to be going through a genetic bottleneck, that is, an extinction event so severe that only a handful of people survived. Moreover, one would expect increased radiation to trigger these mutations. Typhon's passage and the comet impacts at that time would have damaged the ozone layer and disturbed solar activity with the known huge influx of cosmic dust 5,000 years ago. Without normal ozone protection and with increased solar flares, earth should have experienced more mutations than normal.

The DNA-indicated near-extinction event just before 3105 B.C. perfectly lines up with Typhon's passage. This confirms that Typhon is a very dangerous object.

We know Typhon's debris field is returning now. It would also be wise to find out when Typhon's twin, the giant object T-2, is going to make its next close approach to the earth. We may be able to reconstruct more information about Typhon-2's cycle by looking at the close-approach pattern it has maintained over the past 600-million years. If it makes its closest approaches every 65-million years, it is due soon. We must assume it is on a regular orbit that divides evenly into around 65-million years.

We know the 11,592-year cycle does not coincide with Typhon's 5,125-year orbit, and accordingly, would seem to relate to T-2's orbit. Dividing those 11,592 years into 65-million years would produce 5,607.3 T-2 orbits.

The last event on this cycle was almost 11,592 years ago, at the time of the global sea-level rise and the deluge events. So, the cycle should be close to exact, not off by a fraction. We must be near one of T-2's extreme close-approach passages, given that we are about 65-million years removed from the last great extinction. We should not assume it is yet future to us, however. It may be that one of the last few passages of T-2 was the extreme one, and that the next 65-million-year extinction event (that is, the next one after the dinosaur extinction event) is now safely behind us.

Is there evidence for an especially severe extinction in recent time?

Yes. The very last one and its predecessor were both extremely severe, perhaps among the most severe ever. Let's look at the Biblical records:

If we take the Bible at its word, only eight people survived the most recent event, arguably 11,592 years ago, but everyone was sealed inside the ark, unable to observe the passage of T-2 as it came by. It is noteworthy that the text says no animals at all would have survived if Noah had not rescued them. (Genesis 6:13, 17)

The preceding event, 23,184 years ago, had no more than four survivors, if we credit Adam's reference to his mother and his father in Genesis 2:23-24 as evidence for their survival. There is no indication Adam and Eve could have witnessed the passage of T-2. But perhaps the parents, if they survived, had witnessed something of the destruction.

As we survey these events, the first thing we note is how difficult it would have been for any reports of direct observation of T-2 to have survived down to our time.

If we limit our data to scientific sources, we have seen that the severe climate change and magnetic reversal 23,000 years ago apparently led to a near-extinction of mankind in which the DNA of only one woman survived. And 11,600 years ago, mass-extinction of animals, dramatic climate change, severe volcanism, and a 375-foot global sea-level surge wiped out human settlements world-wide.

Both of these events seem extremely severe. But Jesus warned that the next event will also be a near-total extinction, which must be "cut short," otherwise there would be "no flesh saved alive" (Matthew 24:22). As far as extinction is concerned, then, all three events (including this prophesied one) seem to be near-extinctions.

Daniel said the next one would be worse than any that had occurred, "since there was a nation" (Daniel 12:1). Archaeologists believe nation-states did not exist until about 6,000 years ago. So, all

Daniel may have been saying was that the next event will be worse than anything in the past 6,000 years.

Even if we credit Egypt with a history of 18,000 years, that does not require the coming event to be as bad as 65-million years ago. It would only mean that it will be as bad or worse than the Deluge of Noah, some 11,600 years ago. And again, Jesus said this next event would be, "as it was in the days of Noah" (Matthew 24:37).

However, Jesus also stated that "then there shall be great tribulation, such as was not since the beginning of the world until this time, no, nor ever shall be again" (Matthew 24:21). Of this event, Jesus said it would be a total extinction of all flesh if it is not cut short (Matthew 24:22).

That indicates the most extreme passage of T-2 is just ahead. Also, Jesus and the Apostles emphasized an aspect of the coming cataclysm, a specific global calamity that was not part of the Deluge when T-2 came by 11,592-years ago: They warned about fire being "cast on the earth." (Luke 12:49, cf. II Peter 3:10, Revelation 8:5, 16:8). Fireballs are already falling, and fire is definitely typical of Typhon, T-2's primordial twin. As the hard science now reveals, cataclysmic cosmic fire has indeed returned to judge the earth.

Typhon Bumps in the Night

The bright flashes and loud noises being reported worldwide are being described in "apocalyptic" terms, even by non-believers. There appear to be two causes for the kinds of phenomena that the witnesses describe, some of which have been recorded, and which recordings I have now seen and heard. I also have some personal experiences to cite.

The first source for some of the loud noises, as well as the bright flashes of light in the sky, is probably the growing rain of debris from Typhon. These are exactly the sort of thing we should expect to occur as various kind of material rain down from space.

The second source may be deep earth groaning noises and possibly electrostatic light similar to heat lightning that could result from the infernal shifting of earth's mantle and crust as the solar wind fluctuates. Typhon is increasingly disturbing the sun, which changes the tenor of the solar wind. The magnetic field of the earth shifts due to changes in the solar wind, and that affects the torque on the rotating core and mantle. The end result can be a series of loud groaning and trumpet-like wails as the crust and mantle adjust. This may generate weird light effects. Increased quake and sink-hole activity are also related. I have warned about sink-holes for years, and they are increasing rapidly.

Personally, I have also heard strange sounds. One, which I heard right after I had finished recording an audio file, sounded like some heavy object falling down on the far wall. However, I checked the whole house and found nothing amiss. The second sounded like a very loud explosion about 10 p.m. recently, but no one else seems to have heard it.

These are consistent with odd sounds others report. The explosion could have been a fireball breakup, masked by television audio for most people. The other noise sounded as if it were in the room, directionally right in front of me, yet nothing had fallen. A subterranean mantle phenomenon might account for it. Sound waves in rock may propagate unevenly, being heard by those in just the right spot, but not by others nearby.

Chapter Bibliography

Beatty, J. Kelly, and Alan M. MacRobert. 2001. "Fountains of Chondrules from the Sun's Cloudy Birth." *Sky and Telescope*, October: 18-19.

Bower, Bruce. 2009. "Early Brazilians unveil African look." *Science News*, July 1: 212.

Brown, Michael H. 1990. *The Search for Eve*. New York: Harper and Row.

Carlisle, David Brez. 1995. *Dinosaurs, Diamonds, and Things from Outer Space*. Redwood City, California: Stanford University Press.

Choi, Charles Q. 2009. "Neighborhood Darkness." *Scientific American*, January: 24-25.

Clube, Victor, and Bill Napier. 1982. *The Cosmic Serpent*. New York: Univere Books.

Fecht, Sarah. 2011. "MESSENGER Spacecraft Resolves Some Mercury Mysteries and Creates New Ones." *Scientific American*, March: 34-39. Accessed September 2, 2018. https://www.scientificamerican.com/gallery/messenger-spacecraft-resolves-some-mercury-mysteries-and-creates-new-ones/.

Felix, Robert W. 1997. *Not by Fire but by Ice: Discover What Killed the Dinosaurs...and Why It Could Soon Kill Us*. Bellevue, Washington: Sugarhouse Publications.

Ferris, Timothy. 1997. "The Moon's Big Splash." *Natural History*, March: 12-13.

Goldberg, Mark S., Amanda J. Wheeler, Richard T. Burnett, Nancy E. Mayo, Marie-France Valois, James M. Brophy, and Nadia Giannetti. 2014. "Physiological and perceived health effects from daily changes in air pollution and weather among persons with heart failure: A panel study." *Journal of Exposure Science and Evirmental Epidemiology*, June 18: 187-199.

Gribbin, John, and Mary Gribbin. 1996. *Fire on Earth: Doomsday, Dinosaurs, and Humankind*. New York: St. Martin's Press.

Kiefer, Walter S. 2008. "Forming the Martian Great Divide." *Nature*, June 26: 1191-1192.

Lewis, John S. 1996. *Rain of Iron and Ice: The Very Real Threat of Comet and Asteroid Bombardment.* New York, New York: Basic Books.

Mann, Charles C. 2011. "The Birth of Religion." *National Geographic,* June: 42-43.

Milankovitch, Milutin. 1972, 1930. "Mathematische Klimalehre und Astronomische Theorie der Klimaschwankungen." *In Handbuch der Klimatologie,* by Milutin Milankovitch. Berlin, Prussia, Germany: Teil A. von Gebrüder Borntraeger.

National Geographic Society. 1967. "Indian Ocean Floor (map)." *National Geographic,* October: Map Insert.

Science News. 2009. "Astronomy Research Notes." June 30: 352.

Sky and Telescope. 2001. July: 44-51.

Sky and Telescope. 2001. September: 18-19.

Svitil, Kathy A. 2000. "Soccer Balls from Space." *Discover ,* June: 88.

Tennessen, Jacob, Abigail Bigham, Timothy D. O'Connor, and Joshua M. Akey. 2012. "Evolution and Functional Impact of Rare Coding Variation from Deep Sequencing of Human Exomes." *Science,* May: 64-69.

Trujillo, Chadwick A. 2003. "Discovering the Edge of the Solar System: Recent Discoveries Suggest that Planets Larger than Pluto may exist in the outer reaches of our solar system." *American Scientist,* September-October: 424-431. http://www.jstor.org/stable/27858273.

Turner, Michael S. 2009. "The Origin of the Universe." *Scientific American,* September: 72.

van Flandern, Tom. 1993. *Dark Matter, Missing Planets and New Comets.* Berkeley, California: North Atlantic Book.

Velasquez-Manoff, Moises. 2007. "Linguists seek a time when we spoke as one." *Christian Science Monitor,* July 19: 13.

Velikovsky, Immanuel. 1950, 1967. *Worlds in Collision.* New York: Dell Books.

Weed, William Speed. 2001. "Chasing a Comet." *Astronomy,* September: 20.

Chapter 9

The Ends of this Age

For thousands of years, mankind has passed on its memories of past cataclysms as if traumatically-driven to keep repeating tales of destruction. Each culture has handed down its own precious recollections of devastated worlds and the few who had survived them. The stories may vary, but the legends of doomed ages have haunted us all.

Every culture recalls "The Flood," the chilling inundation of torrential water and rising seas, from which a handful huddle in a boat. In the end, they find themselves alone on the wide, muddy earth. They are able to save animals from the former age and start anew. Only these few survivors preserved the slim record of their drowned world.

Most peoples also recall a great burning by cosmic fire. In this horror, flaming orbs rained from heaven, torching the highlands, destroying forests and croplands alike. People could only watch helplessly as their land was devoured by "The Fire." Mankind fled to caves and hid in the earth until the scorching terror burnt itself out.

They numbered the upheavals, and found their cycles, for they recurred over long periods. Depending on the starting point, most reckoned us now in the fifth or seventh age. (Velikovsky 1950, 1967, pp. 46-52) They had also been told these nightmares would end, and the cycles would cease, in fiery doom, when the entire planet would melt, and all things would be dissolved (II Peter 3:1-13): (Velikovsky 1950, 1967, pp. 46-47, 52)

> ...by the word of God, the heavens of old and the earth then standing out of the water (seemingly safe above sea-level) ...being overflowed with water, perished. But the heavens and the earth which now are, by the same word, are kept in store, reserved unto Fire... in which the heavens shall pass away with a rushing noise (a cosmic impact?) ...the elements shall melt with fervent heat; the earth also and the works on it shall be burnt up...all these things shall be dissolved... the heavens being on fire shall be dissolved and the elements shall melt... (II Peter 3:5-7, 10-12)

How close is this ultimate event? Most who knew of it seem to have expected seven or nine ages altogether, counting either from the Deluge or from the time of the first couple, as in the Adam and Eve story. Thus, two ages are yet to come, and then the earth itself shall be melted, dissolved into its very elements. (Velikovsky 1950, 1967, p. 52)

What causes this terminal event?

Immanuel Velikovsky, interpreting Philo and the epics of mankind, thought that the earth would collide with another planet. (Velikovsky 1950, 1967, p. 47) This would not be just another near-miss where two titans slip by each other in the night, but rather, a horrendous collision in which earth and another planet would merge into one great fiery mass of molten rock. Out of this utter inferno would be born a new earth and a new heaven:

> We, according to His promise (Revelation 21), look for new heavens and a new earth in which dwells Righteousness (a title of Christ). (II Peter 3:13)

With what planet might the earth collide?

Typhon and Typhon-2, pieces of an exploded planet, two chunks of its core, continue to orbit like great derelict vessels, circling the dark heaven, periodically coming perilously close to earth. Of these, Typhon may immolate itself in a final Great Conflagration.

Until then, we are told, earth will persist. The current age, like the others before it, will pass through the painful cosmic gauntlet of destruction, but mankind and the planet will not utterly perish... this time.

The day of final destruction has drawn ever closer with each new age. Geneticists tell us mankind went through a population bottleneck of a couple thousand survivors 70,000 years ago, but only a single family 23,000 years ago mothered all of mankind outside of Africa (and much of Africa as well). (Wells 2006) The world's flood traditions tell us no one would have survived the Deluge without divine intervention. Jesus warned that this age right now is facing the total extinction of all life, saying, "no flesh would be saved alive" if divine intervention had not already been taken to "shorten the days":

> And except the Lord had shortened those days (as if this had already occurred prior to the time when Jesus spoke), no flesh would be saved alive; but for the sake of the elect whom He has chosen, He has (already) shortened the days. (Mark 13:20; but *cf.* Matthew 24:22: future tense)

What does Jesus mean by "shortened?" The Greek word being translated "shortened" (Strong's # 2856 κολοβόω *koloboō*) means to cut each individual item in length. In other words, the literal meaning is to reduce the length of each day by several hours. The net effect would be to reduce the total time, as a fixed number of days, which seems to be how it is to be reckoned, according to Revelation 11, where the final period is given as 1,260 days.

We have a confirmation of this interpretation in Revelation, where we are told specifically that the length of each day will be cut short by a third: "The day (each one) shone not for a third of it and the night (each one) likewise." (Revelation 8:12)

Shortening a 24-hour day by one-third requires a reduction of eight hours out of the length of each day. To do that, the earth must have its rotation sped up by 50%. That may not sound right at first, but it is easier to understand if you imagine earth spinning in half the time, once every twelve hours. To rotate around in half the time, the earth would have to spin twice as fast. But spinning 50% faster would only cut the day by a third.

But, how can the days have already been shortened? What has God already done that will, in the future, cause each of the last days to be shortened?

To make the earth rotate faster, an outside force must be exerted. More energy has to be added to move the mass of the earth around at an accelerated rate. Something very large must come alongside the earth, moving fast enough to spin up its rotation.

Accomplishing this requires that the entire earth must be "grasped" by a gigantic force and then spun up by 500 miles per hour at the equator. The force must be gravity, and that has to come from a large massive body, that is, a planet-sized mass. This planetary body must be moving past the earth in the same direction as earth spins, and at least 500 miles per hour faster than earth's current 1,000 miles per hour equatorial rotation. Moreover, this massive object must be moving essentially along the path of our equator, which means it must orbit the sun well within 23 degrees of the ecliptic, the plane of the earth's own orbit.

These are very restrictive conditions. We can describe this object as a planet in an orbit that comes very near the earth and its equator yet is moving significantly faster than the earth does. That combination requires a planet from outside earth's orbit but coming in past the earth briefly at high velocity. This describes an elongated or comet-like orbit.

In other words, the prophecies in question require the existence of a planetary body big enough to grasp the entire rotating earth with its gravity and spin it up faster as it streaks past our world fairly close to the equator. We have seen from earlier biblical passages that this event takes place near the Summer Solstice (Matthew 24:32-34, compare this with the Winter Solstice event, possibly three-and-a-half years later: Revelation 11:3, 10).

One unique thing about the Summer Solstice (like the Winter Solstice) is that an object passing the earth then could parallel the equator while pacing the earth's orbit in the ecliptic plane. The solstices are the only two times of the year when this is possible.

All these things fit Typhon perfectly.

Jesus indicates God has already intervened to cause something (*i.e.* Typhon) to come close enough to the earth to spin up its rotation and shorten the length of our days.

When did God do this?

The earth's orbit has changed. We previously presented evidence the earth used to have a 360-day orbit. (Velikovsky 1950, 1967, pp. 188*ff*, 333*ff*) This is not hard to understand if one imagines giant Jupiter's 4,332-day orbit (12 x 361) acting upon the earth over many billions of years. It would gradually nudge the earth into some kind of synchronization with Jupiter. (Gribbin and Gribbin 1996, pp. 142-143 note 2) We see this resonance effect at work on thousands of objects in the asteroid belt (see chart in *Sky & Telescope*, July, 2001, p. 46).

When earth had a 360-day orbit, it would have synchronized exactly with Jupiter's 4,332-day orbit every 360 years, with both planets returning back to the same locations relative to one another and the galactic heavens. But that perfect pattern is now gone. Our new orbit has no resonance at all, which implies our current orbit is a recent change.

Immanuel Velikovsky (although he made errors on some issues) was on the right track when he said earth's interactions with other planets had changed its orbit. He showed the earth had had numerous encounters with Mars until about 687 B.C. (Velikovsky 1950, 1967, pp. 118*ff*, 333-361) As a test of his ideas, Velikovsky predicted Mars would be found to have had an impact with another planet, that it would be covered with craters and other signs of widespread destruction, that it would show signs of massive volcanism, and that it would have suffered a sudden change in its axis of rotation. (Velikovsky 1950, 1967, pp. 362*ff, et al.*) In all these details, considered improbable at the time, he was proven right. Only the recent timing and specific cause of these things are still in dispute. As with all of his correct predictions, Velikovsky was dismissed as simply benefiting from lucky coincidence.

But by the 1970s, creating computer models of planetary orbits became possible. We could essentially "put the solar system in a box" and watch it change over long periods of time by running high-speed computer simulations (Ed. Note: Anyone can now download a freeware program called "Stellarium" and do this for themselves). We now know that Mercury has the most unstable planetary orbit, followed by Mars and then the earth itself. Astronomers, of course, loathe to admit the possibility of any recent change in the earth's orbit. But they are far more flexible when it comes to things like asteroids and comets. These they consider utterly unpredictable, except to say they are all extremely unlikely to remain in their current orbits. In fact, they find all the solar system's smaller objects in a state of "chaos." (Gribbin and Gribbin 1996, pp. 119-143)

Considering the stabilizing effects of the gas giant planets over billions of years, we should seek the cause of this chaos. (Gribbin and Gribbin 1996, pp. 142-143 note 2) The Kuiper Belt has been disturbed by at least two earth-sized (or larger) objects over a long course of time, given its level of disarray and the long-period orbits of KBOs. (Trujillo 2003) It took many close passages of massive bodies to push these thousands of objects around into the orbits where we now find them; so, this process required billions of years. Why should these same two large interlopers not also have disturbed the rest of the solar system?

This chaotic state of comets and asteroids reinforces our case for the two halves of the exploded core of Saturn II. That is, the evidence shows that Typhon and Typhon-2 (T-2) have been traversing our solar system in comet-like orbits for eons of time, making a chaotic mess of everything they "touch" with their passing gravitational fields.

If these two behemoths, over billions of years, have come close enough to earth and the sun to affect us, then they must also have passed relatively light-weight Mars often enough to have repeatedly moved Mars around, causing it to change its orbit into one that must have occasionally threatened the earth. Likewise, they could also have perturbed the orbit of Mercury, bringing it into contention with the earth and with Mars.

Thus, the chaos in the solar system that Velikovsky saw in ancient records can now be explained: Typhon and Typhon-2 either contributed to it, or else they caused it entirely.

This brings us back to Mars and its role as the leading suspect in changing earth's orbit. In a murder case, we would suspect the last person to be seen with the victim. But, in the case of orbital change, the force needed to move the earth requires a planet-sized body. Mars was the last planet reported near the earth, based on the research on this issue by Velikovsky and supplemented by others. (Patten 1988, note especially p. 63 note 6)

So, it appears that Typhon and/or T-2 disturbed Mars, and then Mars, after several passes, moved the earth into its current 365.242216-day orbit, out-of-sync with Jupiter. The process may have involved multiple nudgings and near approaches. Because Jupiter has not had time to shift the earth into any discernable resonance with its own orbit, we must infer that the Mars passages occurred recently, as Velikovsky's sources insisted.

We have argued from the massive impact basin on each these two planets that Mercury collided with Mars about the time of its last close approach (c. 687 B.C.), moving these two planets into their current unstable, but more distant, orbits. (Gribbin and Gribbin 1996, p. 143 note 3)

One of the names that biblical prophets gave Mars was Tyre. (Turner and Coulter 2001, "Tyr" p. 480) (At another level of meaning, this Tyre also applies to the history of Typhon itself.) Ezekiel says this about Tyre:

> You were perfect in your ways (orbits) from the day
> you were created until iniquity (instability) was
> found in you, by the multiplicity of your travels

(*Strong's* #7404 הלכר, *cf.* #7402 לכר). They have filled the midst of you with violent-motions, and you strayed (out of orbit). So, I will destroy you... I will hurl you at the earth... I will bring forth fire out of the midst of you; it shall devour you, and I will reduce you to ashes upon the earth... and you (Mars and Typhon) will be no more forever. (Ezekiel 28:12, 15-19, author interpretation of Hebrew)

It appears Mars is to be torn asunder during a future close approach to the earth. Mars/Tyre is reckoned Satan's planet. (Turner and Coulter 2001, "Tyr" p. 480) Revelation agrees that the planet of Satan's angels will be destroyed at the end of this age: "Neither will their place (Mars) be found any more in heaven." (*cf.* Revelation 12:8). This suggests Mars is again to be moved out of its orbit by Typhon, on its current visit to the inner solar system. Even if Typhon were to disturb Mars first, Typhon would be on its way past earth before Mars could spiral in toward us. Therefore, any such passage of Mars by the earth would follow Typhon's.

If two gigantic, planet-sized cosmic intruders are advancing toward the earth and are going to be disturbing Mars and the sun at the close of this age, then we will be facing a series of disasters, not just one big event. Accordingly, this age ends with a succession of cosmic cataclysms. Each could be misinterpreted as the final "end," only to be followed by an even worse destruction, until little of the modern world we now know may be left.

What is the likely sequence of these age-ending events?

The first is already happening. As reported, there is a geometrically increasing number of fireballs and clusters of fireballs bombarding the earth. They appear to be igniting forest fires at higher elevations. Our previous fireball chart is already falling behind as an ever-escalating number of fireballs plummet from the sky each month.

We saw previously that ancient traditions from all around the world, not just the Hebrew prophets, expected this age to end violently, in a heavenly upheaval or a cosmic conflagration. There were warnings given by the Chinese, the Japanese, the Polynesians, the Hawaiians, the natives of Australia and New Zealand, the Indonesians, the Hindus, the Buddhists, the Persians, the Egyptians, the Nigerians, the Arabians, the Phoenicians, the Etruscans, the Romans, the Greeks, the Babylonians, the Assyrians, the Finnish people, the Norse, the Icelandic people, the Mayans, the Inca, various native tribes, and, of course, the Bible. (Velikovsky 1950, 1967, pp. 46-52*ff* and also the book's index)

More and more people are reporting clusters of these fireballs, a sign of cometary debris that easily fragments into a cluster of fiery objects when it enters our atmosphere. Increasing numbers of fireballs and objects in each cluster will cause more fires, not only in high country, but at lower levels as well. Downwind of burning forests in America are cattle-grazing and agricultural regions. The nightmare of America is not fire in the mountains, but prairie fires! Prairie lands produce the corn, wheat and soybeans, the cattle and eggs, along with the oil and gas America and the rest of the world need to survive.

Biblical prophets went into detail about these upheavals. Isaiah warns the fire will wipe out agriculture as God's "outstretched hand" (Typhon's debris field) arrives:

> For wickedness burns as the fire (= sacrificial fire of the Altar should atone for sin. God declares it unfit; instead, He lets it burn sinners and their land (Isaiah 1:4, 7, 10-31, 9:6-11:16)), it shall devour the briers and thorns (even tumbleweeds of prairies and cactus of deserts), and shall kindle in the thickets of the forests (fire begins in forests, but spreads* to prairies), and they shall ascend up (back to God, as if sacrificed upon the Altar), lifting (them) up as smoke. Through the wrath of the Lord of Hosts is the land darkened (by smoke of burning crops),[1]

---

[1]The prophecy is to a rich, sinful, hypocritical end-time "Christian" land, hated by Moslems, with great forests upwind of its farms and large population. (thus, not the Middle East, see Isaiah 9:6-11:16)

and the people shall be (sacrificed) like the fuel of the fire; no man shall spare his (own) brother. And he shall snatch with the right hand and (still) be hungry, and they shall eat (even) with the left hand (Middle Eastern people would never do this), and they shall not be satisfied. They shall eat every man the flesh of his own arm... For all this, His hand is stretched out still (Typhon's stretched-out debris field continues to stream past the earth, but Typhon itself is yet to come). (Isaiah 9:18-20)

John in Revelation says that just prior to the Tribulation (as Typhon approaches), the earth is pummeled by a fiery rain of cosmic hailstones (the fireballs). They burn up one-third of the earth's forests and all of its grass:

The angel took the censer and filled it with the fire from the Altar (=Typhon) and cast it on the earth. There was thunder, lightning, and an earthquake (big object hits?) ...a hail of stones, fire mingled with blood (blood-red fireballs), and they were cast upon the earth, and the third part of the trees were burned up and all green grass burned up. (Revelation 8:5, 7)

Wheat and other grains are grasses. So, taken literally, all the world's grain could ultimately be destroyed. The consequences would be grim:

And power was given to (Death and Hades) over a quarter of the earth to kill with sword, and with hunger, and with death... (Revelation 6:8)

So, the Typhon fires lead to famines and wars. Disease will inevitably follow:

And great earthquakes shall be in various places, and famines, and pestilences, and fearful sights and... signs... from heaven. (Luke 21:11 cf. Matthew 24:7, Mark 13:8)

Note the earthquakes... The added component that Typhon brings to the situation is its mass, the tug of its gravity. When the pull of its massive bulk begins to twist the continents around and slide them about, and when the upper crust of the planet gets locked onto Typhon, everything will be wrenched out of its current location. Mountains will suddenly subside, and valleys will heave upward as the continents slide over the basins and ridges of the seafloor.

Typhon's passage will surely cause earthquakes on a scale humanity today cannot comprehend. Nothing will remain as it is. Every mountain will be affected and all the islands as well. By the time T-2 has also come by, no city on earth will remain standing:

> ...and there was a great earthquake, such as was not since men were upon the earth, so mighty an earthquake, and so great... the cities of the nations fell... And every island fled away, and the mountains (having sunk into the sliding continents) were not found. (Revelation 16:20)

As Typhon, Mars, and T-2 each approach, their gravity will shake the heavens: The sun will flare up, and the lunar orbit and rotation will change. The axis of the earth will stagger and shift the heavenly vault of stars, and the oceans will heave and threaten to sweep across the continents, as the bombardment of fireballs escalates to a climax:

> There will be signs in the sun and the moon and the stars, and upon the earth, with distress of nations, with perplexity, the sea and the waves roaring. Men's hearts will be failing because of fear (= *Phobos*, one of the two moons of Mars), watching for those things which are coming upon the earth, for the powers of heaven shall be shaken. (Luke 21:25-26)

The solar system will be disturbed by the passages of Typhon and T-2:

There was a great earthquake and the sun turned black as sackcloth of hair and the moon turned blood-red, and the stars of heaven fell toward earth... shaken by a mighty wind. And the heaven departed as a scroll when it is rolled together, and every mountain and island were moved out of their (previous) locations... (Revelation 6:12-14)

The earth's satellites and space probes are eventually swept away:

Though they climb up to heaven, from there, will I bring them down. (Amos 9:2)

The rich and powerful will cower in terror as the cataclysms escalate:

And the stars of heaven fell unto the earth, even as a fig tree casts her untimely figs, when she is shaken by a mighty wind. And the heavens departed like a scroll when it is rolled together... And the kings of the earth, and the great men, and the rich men, and the chief captains and mighty men, and every slave and feed-man, hid himself in the caves... of the mountains, saying, Rocks (meteorites) are falling on us! Hide us from the face of Him who sits upon the Throne! (Typhon can appear throne-like from some angles) (Revelation 6:15)

The appearance of Typhon (and T-2) changes as it passes the earth. The sun shines first on one side of the tumbling, misshapen, damaged half-dome orb and its long, distended debris field. Then, as it moves by the earth, sunlight reflects off its odd shape at different angles, greatly modifying how it looks. (Velikovsky 1950, 1967, pp. 266-272)

What appears like a bowl or throne-shape upon a near approach may at another time look like an outstretched hand (arm) holding a sword. (Velikovsky 1950, 1967, pp. 266-272) The passage quoted above in Revelation 6 is itself alluding to Isaiah 34, where the appearance is sword-like:

Come near, ye nations, to hear, and hearken, ye peoples: Let the earth hear, and all that is therein; the world, and all (the living: Gen. 1:24) things that come forth from it! (The entire planet and every living thing are afflicted.) For the indignation of the Lord is (poured) upon all nations; and fury upon all their armies. He (T-2 at Armageddon, will) have utterly destroyed them; He has delivered them to slaughter... the mountains shall be melted (as if flowing) with their blood. And all the host of heaven shall be disturbed, and the heavens shall be rolled together like a scroll. And all their host shall fall, as the leaf falls off from the vine and like a fig falling from the fig tree, for My SWORD shall be bathed in heaven. Behold! It shall come down upon Edom ("Red"), and upon the people of My Curse, for Judgment! The SWORD OF THE LORD is filled with blood ... For the Day of the Lord's vengeance, and the year of recompense for the struggle over Zion. And the land (of Edom) shall become burning pitch (petroleum). It shall not be extinguished night nor day; its smoke shall ascend forever..." (Isaiah 34:1-10)

All elements of the Typhon and later T-2 (at Armageddon) passages are here in one text: The blood-red "Sword" in heaven. The disordering of the orbits of the planets. The sudden shift of the heavenly vault as the stars as a group appear to plunge downward toward the earth when the continents slide around. The destruction of the nations and the futility of their military power when confronted by a cosmic body of planetary mass. And the falling fire and volatile petroleum liquids in the comets cast upon the burning, smoking earth when God uses Typhon and T-2 to judge nations at the close of the age.

Typhon's "outstretched arm" is also seen as a harvesting scythe or Sickle of Death. It is the darkly-shrouded and harrowing Grim Reaper = Typhon) (Patten 1988, pp. 72-77)

And I looked and, behold, a white cloud, and upon the (debris) cloud sat one like a Son of Man, on his head a golden crown (Typhon's orb) and in his (outstretched) hand, a sharp sickle. And another angel came out of the Temple (in heaven), crying with a loud voice to him who sat on the cloud, "Thrust in thy sickle, and reap, for the time is come for thee to reap, for the harvest of the earth is dry (literal Greek, i.e. a drought = Isaiah 9)." And he that sat on the cloud thrust in his sickle on the earth, and the earth was reaped. And another angel came out of the Temple which is in heaven; he also having a sharp sickle (= a second sickle in heaven, in this case, the "Blue" Typhon-2's hailstones). And another angel came out of the Altar that has authority over the fire (the Red Typhon's fireballs as in Isaiah 9), and he called with a loud voice to him having the sharp sickle, "Put in thy sharp ('blue') sickle and gather up the clusters of the earth, because her grapes are fully ripe. And the angel put forth his sickle in the earth and gathered the vine of the earth and cast it into the Great Winepress (hailstones) of the Wrath of God." (Revelation 14:14-20)

Here is a plain reference to two separate events that occur in sequence at the end of the age: First, the fiery red "sickle" Typhon, which has power over "the fire" of the Altar cast upon the earth, the Fireballs of Typhon-1, the cometary debris field of the bloody-red object that is already casting its flaming torches upon our planet. This "Grim Reaper," has power to kill a quarter of the earth with its "sword." (see Revelation 6).

Later comes a second "sickle," honed to a shining blue sheen. This later Grim Reaper, the blue or Typhon-2 sickle, is followed by the great battle at Armageddon, where all the armies are crushed by hailstones and the global dictators are defeated by a heavenly host.[2]

---

[2]Revelation 9:13-21, 16:12-21, 19:11-21, Isaiah 24:1-23, 34:1-12, Job 38:22-23, *etc.*

This seemingly odd detail, first a red, then a blue, comet-like object that come at the end of the age, is not unique to the Bible. Native American tribes hold similar views. For example, some tribes expect two successive comet-like stars that will usher in the end of the age. They say one is a red comet-star and one is a blue comet-star. They must be big and close enough to shake the earth, for we are told they will "dance" in the sky.

Multitudes die, although not directly from Typhon and Typhon-2. Most fatalities result from secondary effects. The people die primarily from wars, famines, plagues, and other man-made evils resulting from the refusal of leaders to believe the prophecies, heed the warnings, and prepare the world for what is coming.[3]

Repentance could, of course, be a consequence of someone believing prophecies from God. Yet even without any faith or repentance, as we have seen, the evidence that Typhon and T-2 not only exist, but will again approach the earth, can be independently derived from scientific data. One need not resort to the Bible at all in order to anticipate these cataclysms. This is a testimony to the mercy of God, in spite of His fervent desire that His people repent of their deeds, especially our treatment of widows, orphans, and the poor (Isaiah 10:2). He has nevertheless allowed for the non-biblical discovery of the coming cataclysms, even for those who know nothing at all about His prophecies (Romans 1:16-2:7).

Moreover, the gradual escalation of fireballs that has been going on since 2005 (if not before) has given people time to see with their own eyes the visible testimony and warning from heaven of something very dangerous coming upon the earth, something that is obviously capable of great destruction.[4]

Lest there be any doubt that the divine goal is to protect people from this fiery bombardment (as if endless prophetic warnings were not sufficient proof), Luke recorded this exchange between Jesus and His disciples:

---

[3]Revelation 6:1-17, 9:1-21, 12:17
[4]Joel 1:1-3:21; especially 1:19, 2:1-20, 3:16

> ...the Samaritans... did not receive Him (Samaria (as if America) is threatened with "the fire" in the Isaiah 9 prophecy) ...And when His disciples James and John saw this, they said, "Lord, do you wish that we should command fire to come down from heaven and consume them?" But He turned and rebuked them and said: "You do not know of what kind of spirit you are! For the Son of Man is not come to destroy men's lives, but to save them!" (Luke 9:52-56)

Note that Jesus makes clear that His goal is not only to save men's souls, but to protect them from bodily destruction as well, specifically including fire out of heaven. It is clear that God does not want us to perish from the cosmic fire that is coming. (II Peter 3:9)

Why is God going to "resurface" the earth? To cleanse it of pollution, radioactivity, bio-weapons, cancer- and mutation-causing chemicals, genetically-modified organisms, and sexual, physical, political, religious, and economic exploitation (Isaiah 24:1-6, Revelation 11:18).

The earth's rotation will be accelerated during this tribulation, disturbing the sleep cycles of survivors of Typhon's passage, making even God's people confused about time:

> For the elect's sake, those days shall be shortened (in length) ...the sun shall be darkened and the moon shall not give its light, and the stars shall fall from heaven (fireballs or the sliding continents), and the powers of the heavens shall be shaken (planetary orbits disturbed) ...But of that day and hour knows no man, no, nor the angels of heaven, but only My Father... Watch, therefore, for you know not what hour your Lord comes... Therefore, be you also ready, for in such an hour as you think not, the Son of Man comes... While the Bridegroom delayed, they all slumbered and slept... Watch, therefore, for you know neither the day nor the hour in which the Son of Man comes. (Matthew 24:22, 29, 36, 42, 44ff)

Starving animals will prowl the earth, seeking anyone who might be vulnerable:

> And power was given unto them to kill... with hunger and death (disease) and the wild beasts of the earth. (Revelation 6:8)

As we stated before, like American Sodom and Gomorrah: Las Vegas and Reno, along with Los Angeles, San Francisco, Phoenix, and Salt Lake City, sit on the Pacific Ocean Rift's huge lava caldron. In normal times, these cities are quake-prone. In a time of cosmic calamity, cities like these are certainly at enormous risk of devastation, along with all the coastal cities of the nations that are predicted to fall from quakes and floods.

Now that we have seen the calamities that Typhon brings with it, we should revisit the Egyptian priest's description of Typhon's fiery visitation as it was presented by Plato in his Timaeus dialogue, translating it in more modern terms: (Patten 1988, pp. 72-77)

> Solon, you Greeks are like children, for there is never a Wise Elder who is Greek (the Egyptian priest could just as easily have been rebuking our modern leaders, scholars and scientists who admire the same ignorant Greeks) ...You are all childish in your thinking. There is no old wisdom handed down among you by ancient record (our modern opinion-makers disdain such ancient legends). Nor is there any science that has survived the test of age (modern science produces many unforeseen side-effects: pollution, cancer, lost privacy and freedom, etc.). And, I will tell you the reason for this (naïve ignorance): There have been, and there will be again (when Typhon and T-2 return), many destructions of mankind (destructions of much of humanity), arising out of many causes (including when planets like Mars and Mercury are moved to new orbits). There is a story (so horrific) even you have preserved

(it), that once, long ago, Phaethon (a pre-Flood Typhon name), the son of Helios (the sun), having yoked the horses to the chariot of his father, because he was unable to drive them on the (normal) path of his father (the sun wandered in the sky: This shows earth's rotation is disturbed by Typhon's gravity), burnt up all that was upon the earth (Typhon's fireballs blanketed the planet), and was struck by a thunderbolt (of Zeus: The same story is told by Hesiod, who calls Phaethon, Typhoeus). Now this seems like a myth, but actually describes an approaching (near the earth) of the (cosmic) bodies orbiting past the earth and through the heavens, and a Great Conflagration of things upon the earth repeating over long intervals of time. When this happens, those who live upon the mountains and in dry and high altitudes are more liable to destruction (by fireballs) than those who live by rivers or on sea-coasts... On the other hand, when the gods purge the earth with a Deluge of rain, among you the cattlemen and shepherds in the mountains are the survivors. However, those who live in cities are swept away into the sea. But in this country, neither at such times, nor any other does rain from above fall on our land, but instead tends to surge from below (as a tidal wave). For this reason, the records preserved here are reckoned the oldest... And whatever has happened (of significance) ...has been written down since ancient times and is preserved in our temples. Whereas you and other nations are (a century after the final Mars catastrophe) just now becoming literate and civilized, eventually (= now), after the usual cycle of time, the stream from heaven (Typhon's debris field) descends like a pestilence (fireballs fill the sky like locusts) and leaves only you who are illiterate and uneducated. And so, you begin again like children and know nothing of what happened in ancient times... You remember only one Deluge, but there have been many of them." (Donnelly 1882, 2013, pp. 8-9, revised from Plato's text *Atlantis*)

We can now see clearly that this Phaethon tale is actually about the same large object we have chosen to call "Typhon" (because we have to call it something and we need to distinguish it from its other half, Typhon-2, and from Mars). Like Typhon, Phaethon jars earth with its planet-sized girth, making the sun appear to wander in the sky. At the same time, Phaethon burns the earth with fire, especially at higher elevations and drought-stricken areas. The priest's odd reference to "dry" places implies a drought accompanies Typhon, as does the Revelation 14:15 passage about the fiery red sickle of Typhon coming when the crops are "dry" (that is, during a drought in the summer). The Greeks after this time began to refer to the Typhon events as the "Ekpyrosis" or the Great Summer, when the earth is burned up in a cosmic conflagration. (Velikovsky 1950, 1967, p. 46)

It is noteworthy that the priest chastises the pompous Greeks upon whom so many of our modern "wise men" dote. The failure of Bible translators (other than the late George Ricker Berry) to render the literal Greek as "dry" in the text of Revelation 14:15 is the kind of bogus "wisdom" the priest mocks. Revelation 14:18 uses the correct word for "ripe" a few verses later, obviously to make a contrast with the "dry" or drought-stricken harvest. The translators, however, being ignorant of the very legend that the priest cites, mistakenly edited the vital reference to a "dry" harvest out of the prophecy.

The repeated biblical mention of drought and famine in connection with Typhon's passages must be important. How are they linked? Several possibilities exist. One is that Typhon may somehow disrupt the sun with its dust and debris field, and possibly its gravity and magnetic field. The disturbed sun might then cause the drought.

Or, fireballs that precede Typhon by a few years may directly trigger the drought. Or they may sweep the inter-planetary dust needed for rainfall generation out of our path.

The visits of Typhon deposit dust in the inner solar system every 5,125 years. This dust stimulates rainfall and boosts food production, as it did after 3100 B.C. But the dust is gradually scooped up by the earth and other planets. Today, although some remains, the effect on precipitation has diminished significantly. Modern agriculture and irrigation have offset some of this reduced fertility of the earth, but everyone knows there is a limit to growth as global population outstrips the potential food supply. Increased rain in the food growing regions could certainly make today's harvests look "dry" by comparison.

Most likely, all of these things contribute to the connection between the droughts and Typhon's passages. But, as we can see, none of these things is reversible by human intervention. We cannot keep the fireballs from sweeping the dust out of earth's path, for example. We cannot control solar activity. We cannot make up a global rain deficit with irrigation and desalinization on short notice. Even on a long-term basis, we have not made much progress with such measures, even though governments around the world claim to be seeking to solve the water shortage with desalinization of ocean water.

This reveals another aspect of the problem. The Egyptian priest alluded to rivalry between the rural mountain-dwellers and the seacoast city-dwellers. The rural group provided food for the cities, but the city folk looked down on them. That same divide is apparent today, between religious rural conservatives and agnostic urban liberals.

This rivalry will be affected if the passage of a massive body near the earth shifts the continents out of their places. Typhon is big enough to reshape the earth. And Typhon seems to return at the summer solstice, passing by parallel to the equator.

If this were to happen, the whole planet might remain tilted about the same as it now is. But the outer surface would shift; the bulk of the planet itself (that is, the inner part) could be much less affected.

In any case, based upon what the alternative poles of the earth were during previous shifts, the eastern coast of the United States could end up in a position along the earth's rotational equator.

The crustal shift would cause some colossal tidal waves as it was happening, but the long-term effect would be dramatic climate change. The agnostic liberal eastern United States, if many coastal people survive at all, could become very hot and humid.

On the other hand, Antarctica could be moved 45-degrees north, and that would cause its great ice-sheets to melt. That in turn would raise global sea-level, if the shift does not result in increased ice build-up in conservative rural areas of the western United States.

Very few locales on earth, if any, would keep their current weather. Many of the world's natural ocean harbors would be iced-over or be washed away or shifted inland. If a nation today prospers by its seaports, it could find itself without them after Typhon!

Agriculture would not be practical in many of its current locations. However, farming could begin to flourish in new places to which the rain would have shifted.

The religious rivalry will also manifest in how people view what happens next. Many people expect aliens or angels to show up in such times. Some think aliens that have been hiding in the "shadows" will finally appear, or angels will battle our military. Others expect Nibiru, Planet X, time-travelers, and/or secret Nazis. Revelation says this:

> There appeared another sign in heaven... a great red Dragon... (After John saw Jesus escape the Dragon,) his tail drew down a third of the stars of heaven (cutting short the day by a third when red Typhon's debris tail comes by earth; Typhon was described by the Greek Apollodorus as a fire-spouting blood-red dragon in the heavens who pelted earth with burning rocks; Zeus and Typhon fought over "an adamantine (diamond-glittering) sickle;" Typhon is cast to earth as the mountains "heaved." (Velikovsky

1950, 1967, pp. 65, 93-99) (Patten 1988, p. 74)
...(Then) there was war in heaven, Michael (the
archangel) and his angels fought against the Dragon
and the Dragon and his angels fought back but did
not prevail. Neither was their place (Mars) found
any more in heaven (ashes and rocks of battered
Mars fall to earth). "Woe unto the inhabitants of
the earth and of the sea! For the evil one is come
down to you in great wrath because he realizes his
(expected) time is short!" (Revelation 12:3-4, 7-8,
12, *cf.* Ezekiel 28)

The ancient description of Typhon is the same as in Revelation.
"An adamantine sickle" refers to a diamond-like sickle. (*The
American Heritage Dictionary of the English Language* 2016, p. 14b)
That is, it glittered like diamonds (cf. "glittering" sword or spear in
Deuteronomy 32:41, Habakkuk 3:11). However, Typhon acquired
a cloud of genuine glittering diamond dust from its gouging
encounter with Uranus.

In the aftermath of Typhon's upheavals, we can be sure that a
strong leader will emerge, much like when King Menes took control
of much of the earth after Typhon came by *c.* 5125 years ago:

And the Dragon gave him his power and throne,
and great authority ...And all the world wondered
after (the new ruler): "Who is able to make war with
him?" And there was given him a mouth speaking
great things and blasphemies. And he was granted
power to continue (in his office as the post-Typhon
global ruler) for 42 months." (Revelation 13:2-5)

Unfortunately, we cannot be sure how long "42 months"
is after "the powers of the heavens are shaken." The moon will
be disturbed. Its orbit may be affected. The city of Ephesus was
fanatically devoted to the moon goddess Diana. (Acts 19:24-35)
But we are told that its "candlestick" (actually a "star" (Revelation
1:20)) might get "removed... out of its place" in the last days if it

does not repent. (Revelation 2:5) There seems to be no repentance by Ephesus yet; so, the moon may be likewise "removed...out of its (current) place." If it is "removed" into a solar, rather than earth-bound, orbit, then the moon could take as long as 40 or 50 years to complete "42 months" of orbits.

No wonder Jesus says that even His disciples and the angels would have trouble understanding "the days and hours" of the end times. At the time He spoke to them, the "day" was reckoned by the cycle of the phases of the moon. Indeed, the Jewish and the Muslim calendars are still based upon calculations of that monthly lunar cycle.

And so, we cannot now know exactly how long a global administration will be allowed to rule. Seizing emergency powers to rule by their own executive fiat and martial law, they will seek to impose a global military dictatorship. We are told that they will try to stop dissent and armed resistance:

> And it was given to him (perhaps by some international gathering that votes to have him do certain things) to wage war on the saints and to subjugate them; and power was granted to him over all kindreds, and tongues, and nations. And all that dwell upon the earth shall worship him... (God tells His people to be patient:) "He that leads into captivity, shall go into captivity. He who kills with the sword must be killed with the sword." (That is, if people try to take prisoners, they will themselves be taken prisoner; and if they try to kill the oppressors, they will only get themselves killed.) ...And he (a secondary ruler who takes over when the chief ruler is wounded; this second figure is also called the "False Prophet," a religious leader who deceives people) exercises all the power of the first (ruler, who had received his power from the Red Dragon: also seen as the fireball-casting Typhon) ... even making fire come down out of heaven to earth in the sight of men." (Revelation 13:7-8, 10, 12-13)

The translators usually call these two leaders "Beasts," but the Greek word means "a hunted (or "trapped" = dangerous) animal" (see *Strong's* # 2339 θήρα, #2340 θηρεύω, #2341 θηριομαχέω, #2342 θηρίον). This seems to mean that these two rulers know they are trapped (in a hopeless situation) or doomed (to fail), making them more dangerous. The assassination attempt demonstrates that their job is not safe: There is armed resistance, and prisoner-of-war camps are set up (Revelation 13:10).

The rulers seem to be overwhelmed by the task of restoring order and rebuilding. Cars and trucks appear to be at least partly replaced by horses (Revelation 14:20). Order will be restored, and technology will be rebuilt, because we are told the whole world is able to watch and hear the Two Witnesses prophesy in Jerusalem, and see them crucified on Olivet, "where also their Lord was crucified" (Revelation 11:7-8, Matthew 27:51-54). Also, the entire world "sees their dead bodies" lying in the street (Revelation 11:9-12).

This prophecy of global live television is remarkable, especially given how vast a destruction their world will have undergone. Rebuilding this electronic infrastructure takes years. It is hard to fit this into only 42 months, at least, as we now count them. This implies that the moon's orbit does indeed change. At a minimum, it shows the global ruler compels survivors to work like slaves to rebuild so quickly. After the fire of *A.D.* 64, the Emperor Nero used slaves to rebuild Rome in a very short time. (Tacitus 1977, vol. xv, p. 41*ff*)

Another sign that communications and the electrical grid are largely restored is the creation of a single new global money system:

> And (the "false prophet") causes all, both small and great, both rich and poor, free and slave (note: there are both wealthy people and slaves), to receive (or "present:" the Greek can mean both) a mark (Greek *charagma* means "a scratch-mark" that looks like a row of stakes or lines, hence, a bar-code: see Strong's #5480 χάραγμα and 5482 χάραξ), in their right hand or on the face of them (the Greek has

a dual meaning: It can be a "badge" like a credit card held in the hand or a scratch-mark "stamped" on the "face" of the buyer or the thing being sold, exactly like a modern product bar-code), and (he commands) that no one may buy or sell, unless he had the mark or the name (this Greek word for "name" can mean "authority") of the (ruler), or the number of his name (that is, the "authorized number" stamped on the "face" of the product). (Revelation 13:16-17)

The system being described is one in which products sold have a bar-code on the front of them that looks like a row of lines, with an equivalent number, as a way to buy the product. That perfectly describes, using the limits of the ancient Greek language, our own modern product bar-code system. It is based upon the Universal Product Code, (UPC) in which the row of lines and the number are equivalent in meaning. The mark is "read" by an electronic laser that scans it at the check-out, at the time of purchase. Only when the laser fails to decipher the code does the equivalent number get typed into the computer.

It is astonishing that such a detailed and perfectly accurate prophecy of a modern technological system could be made in ancient Greek. But it shows that there must be time to rebuild the electronic laser industry and manufacture enough equipment to have it installed world-wide. That this electronic system is used to not only track products of the revived economy, but to track people, is also revealing. This system is rebuilt because it enables the ruler to exercise dictatorial power over humanity, possibly to help ration a limited supply of manufactured goods and food.

Laser bar-code purchasing, and live global television require a global Internet. Even ID-theft is implied by the need to "mark" people to verify their identities.

To argue that a 100-year old man (if the Apostle John in *A.D.* 96) or teenager (if John Mark in *A.D.* 50-60) 1,900+ years ago simply imagined all this is unreasonable. There was no technological basis for anyone at the time to conceive of monitoring all global finance to require people use such a system when transacting business. Why would John envision the need to brand people with a number to determine their identity when making purchases, unless there was a way for someone to steal a person's identity? And how could even one person, much less the whole world's population, view events live as they happened in Jerusalem from another country? Moreover, how did John conceive of people making an "image" of the ruler that seemed to be alive and speak? (Revelation 13:15)

The only way to make any sense out of these things is to accept the reality of the visions that John insisted he had had. Somehow, John actually saw the future. He wrote down, as he was able, what he was shown, just as he said (Revelation 1:1-2, 11-20, 4:1, 10:11).

Here's the scary part: John saw our world, it was all our current technology. John saw technologies that will be greatly changed from what he saw, if the world's scientific development were to continue for just a few more decades before Typhon intervenes. That does not give us much time.

If you doubt this, just remember that personal computers were only invented in the late 1970s and that surfing the web with a full-color browser only began in the late 1990s. Wireless global video is newer still. The pace of change is so fast that computers bought only five years ago cannot handle the latest applications. Currently, manufacturers are planning computer "glasses" that would enable the wearer to be "surrounded" by 3-D Internet live video images from the far side of the world, while talking on the phone to one person and texting another.[5]

So, we must be getting close to Typhon's arrival and the upheavals that it brings our world. The growing influx of fireballs also shows Typhon is getting near.

By the way, the prophecy seems to show that some people may still be using an ordinary credit card "in their right hand" instead

[5]Ed. Note: Olaf was unaware of the Nintendo Wii and other video gaming systems that have taken hold in the last few years.

of getting a bar-code mark branded on their forehead (or more likely in the forehead bone via a laser so that the mark is not visible unless scanned by another laser). (*cf.* Revelation 13:17)

Bible interpreters seem unaware the magnetic strip on a credit card is also a form of bar-code, except its number identifies a person rather than a product. If a credit card is the "mark of the Beast," then Europe and America are already doomed. But the text says the branding of one's body is what causes one to get "sores" on the skin. (Revelation 16:2, 11) That is yet another amazing prophecy. We now realize that electronic radiation is capable of causing skin cancer, such as frequent laser scanning to read a mark like this. But, branding with a mark (to prevent ID theft?) will apparently be optional. (Revelation 13:17)

I should point out that people who are spiritually condemned because of the mark are identified as also simultaneously "worshipping" the image of the Beast, not just using a card with the mark.[6] Worshipping of the image is the thing that condemns someone. Simply using cards with bar-codes does not.

The real problem, then, is that the secondary ruler has the authority to order the execution of someone who refuses to worship the image of the prime ruler:

> He has power to give breath to the image (of the ruler), and also cause the image (of the ruler) to speak, and (he can cause) that as many as would not worship the image (of the ruler) be executed. (Revelation 13:15)

Technically, although the text says "he has power" to impose this death penalty on anyone refuses to worship the ruler, it does not say he actually executes "everyone" who refuses to do so. Clearly, the armed insurrection and the existence of a prison camp indicate that not everyone is obeying the rules. More to the point, the whole world will be tuning in to the Internet to watch and listen to the "rebel" leaders in Jerusalem, the Two Witnesses (See Revelation 11).

---

[6] Revelation 15:9-12, 16:2, 19:20, 20:4

That cannot be what the rulers want people doing. Obviously, they are still unable to control the Internet and what the public does with it, much like today.[7]

Thus, in spite of the rulers trying to impose their dictatorship, much of the world ignores some of their new laws. Moreover, entrepreneurs get rich during the recovery. And it appears even dissenting Christians who do not get branded and worship the image survive if they do not revolt and if they keep a low profile (Revelation 13:10; 14:11-12).

The post-Typhon survival of things like the Internet and credit cards shows that, as devastating as Typhon and a Mars passage would be, the prophecies imply civilization as we now know it will not be utterly destroyed until the Armageddon return of Typhon-2. That "twin" of Typhon is now apparently noticeably larger than the surviving remnant of Typhon itself, which was gouged out by its violent encounter with Uranus. So, Typhon-2 would be capable of doing severe damage, and eventually will (Isaiah 34, Revelation 19).

Until then, after Typhon, Mars, and their comets depart, the inner solar system will be filled with fine dust, which will be gathered up by the earth. This influx of dust will stimulate precipitation, and the earth will begin to recover. Typhon's dust will increase rainfall so much the deserts will blossom. It will seem like the Millennium has dawned. Other than believers in the Bible, (Revelation 12:17) most people (except for slaves) will see it as a time of great renewal and growing wealth, albeit for only a few.

A measure of peace and prosperity will develop under the ruler. People will begin to wonder if perhaps he is the long-awaited messianic figure who comes at the end of the age. Some people will be absolutely convinced the Millennium has now begun.

Those who realize that the age has not yet ended and that all the prophecies have not yet been fulfilled will begin to dissent. At

---

[7]*Editor's Note:* Olaf Hage died before the cryptocurrency craze took hold. He would have been shocked at how confidentially, someone could do financial transations all over the world without the use of banks. Blockchain technology has changed how the world transacts business.

some point, the leaders of these skeptics, the Two Witnesses, will begin to preach in Jerusalem. This is said to begin about three-and-a-half years before Typhon-2 arrives (Revelation 11).

The good times will begin to falter as the climate will start to deteriorate. It will be blamed on the Two Witnesses:

> The Two Witnesses... have power to shut heaven, that it not rain in the days of their prophecy and have power over waters to turn them to blood, and smite the earth with all plagues as often as they wish." (Revelation 11:3, 6)

Note that they do not have to do any of these things, but they can whenever they choose. We are given to believe that they do impose some plagues (Revelation 16), but they also seem to be able to lift these plagues as well. We are not told what these miraculous powers entail. It seems God gives them only prophecies that He knows will be fulfilled.

What is the mechanism God uses to fulfill the prophecies of these witnesses? We are told the climate will decline because the sun is disturbed by something:

> And the fourth angel poured out his bowl upon the sun, and power was given him to scorch men with fire. And men were scorched with great heat, and blasphemed the name of God, who has power over these plagues; and they repented not to give Him glory. (Revelation 16:8-9)

The "angels" are called "stars" in Revelation (1:20). Could it be that a "star" (or perhaps some cosmic body that looked like a "star" in John's vision) will come near our sun and "pour out" something into the sun that affects its output?

This may be the debris cloud from the unexpected arrival of Typhon-2, returning for the first time since the Great Deluge that ended the ice age. Typhon-2, not observed at all by Noah's sequestered family, has been long-forgotten on its lengthy 11,592-year journey.

Why would its return cause a great solar drought? Typhon-2's large mass and gravity could affect the sun. And it may also have a powerful magnetic field. But the "poured out" bowl of the angelic "star" suggests the possibility that it is Typhon-2's enormous debris cloud that "pours" into the sun to bring about a flaring of its heat. The over-active sun will evaporate up the seas:

> And the second angel poured out his bowl upon the sea, and it became like the (dried) blood of a dead man, and every living soul in the sea died. And the third angel poured out his bowl upon the rivers and fountains of waters, and they became blood... And the sixth angel poured out his bowl upon the great river Euphrates, and its water was dried up, that the way of the kings of the east might be prepared... to gather the kings of the earth... to the Battle of that great Day of God Almighty... And he gathered them together into a place called in the Hebrew tongue: *Armageddon*. (Revelation 16:3-4, 12, 14, 16)

The solar spectrum will shift out of the visible light into the infrared. The sun will appear dark and the moon will look bloody red:

> ...there was a great earthquake, and the sun became black as sackcloth of hair, and the moon became like blood. (Revelation 6:12)

> And the fifth angel poured out his bowl upon the throne of the Beast, and his (global) kingdom was full of darkness. (Revelation 16:10)

After the sun has evaporated much of the earth's water up into the air, the atmosphere would be super-saturated with moisture. To keep all this water suspended in the air, the sun would have to remain "hot." But this shift into the infrared, which does not penetrate through the atmosphere well, would dramatically cool down the earth.

So, when the sun seems to go dark, it portends an ominous build-up of potential precipitation, and in this case, it will be snow and ice:

> Have you entered into the treasuries of the snow or have you seen the storehouse of the hail that I have reserved against the time of trouble (Tribulation), against the day of battle and war?" (Job 38:22-23)

> And the seventh angel poured out his bowl into the air (which is water-saturated)... and there were... thunders, lightnings, and a great earthquake, such as was not since men were upon the earth, so mighty an earthquake and so great... and the cities of the nations fell... and every island fled away, and the mountains were not found. And there fell upon men a great hail out of heaven, every stone about the weight of a talent (hail about the weight of a small cannonball) ...the plague was exceedingly great." (Revelation 16:17-21)

Notice that the heavy hailstones fall "upon men" on "the day of Battle and War." These heavy, skull-shattering hailstones fall on a great number of men on the day of a great age-ending war. This describes the battle of Armageddon, which is gathering at the very same moment (Revelation 16:16-21). Typhon-2 not only disturbs the sun, but it also pelts the earth with dust, ice and stones of its own, helping to generate even more hail. With this battle, our age finally ends (Revelation 14:19-20, 16:17, 19:15-21, Isaiah 34:1-6).

The Typhon cycles are coming due now. The ancients knew of these great cycles, and they viewed the expected confluence of so many cycles around the same time as a generation of grave danger for all of humanity. Their warnings and prophecies about great earth ages and cycles are directed at our generation. The whole of the ancient world shared a common acceptance of the inevitability of the violent end of this age.

Yet, when people as diverse as Isaiah, Heraclitus, Plato, the apostles of Yahushua (Jesus), the Mayans, Ignatius Donnelley, and Immanuel Velikovsky drew attention to the evidence for cycles of destruction, the "wise men" of their time (and ours) have ridiculed their warnings.

In spite of the ever-growing amount of evidence for these cycles, and in the face of the plain demonstration of cyclical upheavals on this planet, the official response remains one of denial and ridicule. Our intellectual leaders revert to taunts and censorship to silence all who challenge their fictitious view of the distant past. Civilization is ever-rising, they insist. It is like an immortal beast whose long gradual, but unhindered, ascent has steadily carried us up the mount of progress. The journey is ever upward, never down. They claim there was no Deluge, no Conflagration, no cycle of World Destruction.

Yet, the global layers of mega-death refute their faux vision of the undeterred rise of mankind from cave to cosmos. Winnows of corpses once washed blood-red seas and shall again. But it is not God's will that we should perish (II Peter 3:9). His prophets plead with each generation. Yet, refusal to heed the warnings continually traps the imperiled:

> What I say to you, I say to all: Watch! ...look up and lift up your heads ...For, like a trap, it (Typhon) shall come upon all who dwell on the face of the whole earth. Watch... that you may be... worthy to escape all these things that shall come to pass..." (Mark 13:37, Luke 21:26, 35-36)

Chapter Bibliography

Donnelly, Ignatius. 1882, 2013. *Atlantis, The Antediluvian World.* Seattle, WA: CreateSpace Independent Publishing Platform.

Gribbin, John, and Mary Gribbin. 1996. *Fire on Earth: Doomsday, Dinosaurs, and Humankind.* New York: St. Martin's Press.

Houghton, Mifflin, Harcourt. 2016. *The American Heritage Dictionary of the English Language.* 5th. New York, New York: Houghton, Mifflin, Harcourt.

Patten, Donald W. 1988. *Catastrophism and the Old Testament.* Seattle, Washington: Pacific Meridian Publishing.

Tacitus. 1977. *The Annals of Imperial Rome.* Translated by Michael Grant. New York, New York: Penguin Classics.

Trujillo, Chadwick A. 2003. "Discovering the Edge of the Solar System: Recent Discoveries Suggest that Planets Larger than Pluto may exist in the outer reaches of our solar system." *American Scientist,* September-October: 424-431. http://www.jstor.org/stable/27858273.

Turner, Patricia, and Charles Russell Coulter. 2001. *Dictionary of Ancient Deities.* 1st. New York, New York: Oxford University Press.

Velikovsky, Immanuel. 1950, 1967. *Worlds in Collision.* New York: Dell Books.

Wells, Spencer. 2006. Deep Ancestry: Inside the Genographic Project. Washington, D.C.: *National Geographic Society.*

Chapter 10

Mars and Typhon

By the time the Greeks wrote down their myths and the Romans adopted them, the ice age Kingdom of Atlantis had been lost for over 9,000 years. Even Typhon's last visit had been 2,650 years before Plato. The traditions had become muddled and confused. Yet, there was a lot of valuable information scattered among these fables, if one knew how to separate the wheat from the chaff. That required spiritual discernment and the lost keys to Genesis history that the Phiabi line of priests had denied the people of Jesus' day.

Working on these problems for several centuries has allowed us to make progress in deciphering the huge mass of Western Classical and Mideast Mythology. Discoveries of tablets and forgotten sites by archaeologists, the publication of numerous research tools, and even concordances have made the task easier.

But without confidence in the credibility of the Biblical texts and a spiritual faith in the righteous intent of the Creator and His prophets, most researchers eventually go astray, wandering down blind alleys, becoming enamoured of their own carnal theories. Once adopted, these theories become encased in ego-concrete, inviolate.

However, I have learned that discovering errors in one's own theories is the fastest way to make progress. I once thought Jesus had been born in 7 B.C., crucified in *A.D.* 30, and that He wanted us all to become Roman Catholics. I've had to abandon each of these ideas as my research proceeded. Again, and again, I had to face the fact that I was wrong!

For seventeen years before 1997, I thought the only way to "cut short the days" was for a huge asteroid to hit the earth and spin it up; that turned out to be too deadly a scenario. Likewise, from 1997 until 2008, I had been convinced that the primary planetary threat to the earth in the last days would come from Mars. I knew something would jar it out of orbit to threaten the earth, but I assumed that would just be some asteroid or Kuiper Belt comet. I never expected anything as big as Typhon.

Had it not been for God humiliating me over and over for decades as I discovered belief after belief that I held was in error, I doubt I would have so readily embraced the discovery that Typhon was coming and that its was bringing a huge cloud of fireballs.

Yet, the last six years have amply confirmed that the fireballs are escalating in both number and in the rate of increase in their numbers. This requires that a truly massive object is lurking in the midst of the main cluster of debris that causes fireballs. An unimaginably huge sea of cosmic debris stretches out along the orbit of Typhon.

This huge sea of fragments came from an exploding planet four-billion years ago, followed by a collision of the exploded core-halves 600-million years ago, and an impact of one of the two core remnants, Typhon, that toppled Uranus *c.* 3.2 million years ago.

Along the way, Typhon, being a massive object heavier than the earth, has vacuumed up all sorts of odds and ends on its many journeys through space, traversing both the asteroid belt and the realm of Kuiper Belt Objects twice during every orbit, bullying its way through them on its way in and on its way out.

Like a massive cattle drive that covers the land for miles around, Typhon is now herding its vast throng of debris back into the inner solar system, pouring fireballs into the earth's orbit in such profusion that they are becoming scattered all along our path throughout the year.

Thus, the object that exploded over Chelyabinsk, Russia (pictured to the right in the shaded area), in February of 2013, had become diverted into the earth's orbit and became estranged from the main body of debris along Typhon's path.

While most of the fireballs had been falling in late June and around Halloween, the cloud of debris is becoming so large and dense that it is overwhelming the ability of the solar system to keep it herded together. Debris is breaking off the main cloud and is accumulating along earth's orbit, from where it can now fall at any time of the year.

Part of the reason for this growing load of cosmic garbage in our path, however, is that Typhon's orbit itself parallels our own orbit very closely.

This is again confirming our projections, based upon its previous passages as they are recorded in myth and legend. We determined that the massive geological effects that seemed to be associated with Typhon required that it come perilously close to the earth during the time it was making its perihelion passage, closest to the sun.

That is, the innermost portion of Typhon's long orbit tracks along right next to the earth's own orbit for a period of several months before it whips back out into space again.

At least, it used to do that.

But Mars and Mercury were disturbed out of orbit and began to come by earth repeatedly for 2,400+ years after Typhon's last visit, *c.* 3123 B.C. That led to a change in earth's own orbit: We now take approximately 5¼ days longer to orbit the sun than we did before. (Velikovsky 1950, 1967, pp. 333-361) That means we are now orbiting a bit further out much of the time, but we also swing inward, following a more elliptical orbit overall.

The net result is that the earth could now stray right into the path of Typhon. The prophecies do not anticipate such an event any time soon, but they do hint that a collision and total destruction of the earth may come in some future passage, presumably at least 5,000 years or so down the road. This event is suggested by the following passages:

> The heavens and the earth... are reserved for fire against the Day of Judgment and destruction of ungodly men... in which the heavens shall pass away with a great noise, and the elements shall melt with fervent heat; the earth also, and the works in it, shall be burned up... all these things shall be dissolved... We look for new heavens and a new earth... (II Peter 3:7, 10-13)

> I saw a new heaven and a new earth, for the first heaven and the first earth were passed away. (Revelation 21:1)

This comes long after Christ's return, and it is not an immediate problem, but we are warned about it. The value of it is in the recognition that Typhon colliding with the earth is fully capable of fulfilling such prophecies literally. Thus, these provide further confirmation that Typhon's orbit comes very close to the earth and that it is a huge object capable of turning the earth into a molten uninhabitable mass one day.

To be frank, the sun will also expand out and swallow the earth one day billions of years from now, according to solar scientists. Of course, they have been wrong before when it comes to the timing of events. But that would have to be a really big error if it were to happen any time soon.

Even so, the evidence is mounting that Typhon is truly massive and is going to pass terrifyingly close to us. You will not miss it. It will fill the entire sky.

The size of Typhon and its debris field also means that anything lying along its path is probably going to be disturbed by it.

We have already noted the large number of comet impacts occurring on the other planets, including several hitting Jupiter over the past 30 years. (Howell 2018) We can expect this process to escalate as time goes by. The earth is not the only object at risk from Typhon.

The object most likely to be a problem is, of course, Mars. But Mars is only one-ninth the mass of the earth. So, an impact of Typhon debris on Mars is going to have nine times the effect on its orbit as would something hitting the earth. Moreover, Mars is riding around nearly at right angles to the orbit of the debris field. That is, a sideways impact will have a greater tendency to change Mars' orbit than an object hitting the earth, which is moving more or less parallel to the Typhon orbit where it passes us.

Therefore, Mars is uniquely vulnerable to getting hit on its side by Typhon debris and having its orbital shape modified dramatically. And being lighter, Mars has the potential to be hit hard enough to drive it into the path of the earth.

At the moment, I see no reason to abandon the idea that Mars is going to be hit and will come out of its orbit and threaten the earth. In fact, the new fireball data appears to increase the likelihood of that happening. And, the increased estimate for the mass of Typhon also raise a greater chance that it could tug Mars out of orbit as it goes by.

On balance, then, the prospects for a Martian passage by our planet have been significantly enhanced by the escalating number of fireballs. The only questions now seem to be: When will this happen? And what will Mars do to the earth as it comes by?

The first question may be answered shortly. An object is due to pass very close by Mars in October of 2014. It could hit the red planet. The mass of the object may not be very large, but it will provide a baseline for other impacts. We will be able to measure small changes in the Martian orbit if it gets hit significantly.[1] (Phillips 2013) (Watson 2014)

---

[1] *Ed. Note:* After Olaf Hage died, the comet did miss Mars, but the scientists were still glad for the opportunity to learn more about the "Oort Cloud."

Also, the infrared-detection satellite program appears to have been capable of more data collection than at first announced. New information is now coming in that is providing additional discoveries of near-earth objects. These may be Typhon-related.

But I have been wondering if there might be more evidence to be gleaned from studying Typhon's previous encounters with the earth, as described in mythology. This is always a risky business, but that should not deter us when the stakes are this high. We can learn a lot from what the ancients witnessed happening in the skies.

In some of our prior investigations we learned that the battle of Zeus and Typhon was, at least in cosmic terms, a "battle" between Mars and Typhon. Is this a dependable story? We can gauge how reliable such Greek traditions might be by comparing some other Greek myths whose contents closely parallel the Bible.

For example, when Satan is called "that ancient Serpent" (Revelation 12:9), John Mark's Greek uses the term Ophis (*Strong's* #3789 ὄφις). There was already a detailed Greek legend associated with the ancient Serpent Ophion. The Greek Ophion lived in an ancient era that predates the creation of Pelasgus, a version of Adam, being the male of the primeval first couple. (Turner and Coulter 2001, p. 175a)

John Mark's mention of "ancient Ophis" (the different ending is merely a function of Greek grammar) suggests he identified Satan with the ancient Serpent Ophion who entwined and raped Eurynome, whose name means goddess of the North, who gave birth to the planet gods. (Turner and Coulter 2001, p. 175a) She was Egypt's Tefnut ("North goddess") or Tsephon-ette ("North lady"), ancestress (or mother) of Adam and Eve (*cf.* Genesis 41:45, where Joseph is "Zaphenath-paneah"). Does this mean that Ophion was Satan?

Not exactly... The Greek legend may be a tradition about a pre-Adamic woman named "Tsephon-ette" whose story they mixed with the account of Eve being "beguiled" (entwined) by the Serpent. But this story still provides confirmation by an independent pagan Greek source concerning several things alluded to in Genesis: That a Tsephon-ette existed in the pre-Adamic world; that she was an ancestress of Adam and Eve; that Satan beguiled a woman and impregnated her in that primeval epoch, and that Satan was identified as an "ancient Serpent."

In other words, we are finding that the pagan world had numerous bizarre myths based upon things that are described in a much more straight-forward way in Genesis or other parts of the Bible. These pagan traditions occasionally add a few lost details about stories in Genesis, helping us to make sense out of things the Bible mentions in passing and which would otherwise have been overlooked in its text.

The same thing may be true about myths concerning Mars and Typhon. Knowing that stories about Zeus can refer to Mars during the ice age is revealing. For example, we are told that Zeus was the "youngest" of his father's sons and had been hidden in a cave until he could claim his heavenly throne. (Turner and Coulter 2001, p. 522-523)

This implies that Mars may once have been jettisoned farther away from the earth and was so hard to see that it was deemed "hidden in a cave" for a long time, only to suddenly reappear just as it began to "battle" his giant father Cronos and the Titans for heavenly supremacy. (Turner and Coulter 2001, p. 522-523) One of his first battles was with Typhoeus, who is identified with Typhon by some, but is otherwise called Typhon's "father." (Turner and Coulter 2001, p. 480)

It appears that Zeus/Mars fought two successive "battles" with Typhon in two different "generations," roughly 5,000 years apart (that is, one Typhon orbit), when Typhon (as Typhoeus) first shifted Mars out away from the earth, and then later, on Typhon's next orbit, when it moved Mars back in close to the earth.

Since we have already determined that the second event corresponds to the great comet explosion c. 15,000 years ago, the one that flash-melted the corridor in the vast Laurentian ice sheet over Canada, allowing men and mammoth to enter America from Siberia, we may reasonably infer that the first event, involving "Typhoeus" (Typhon's earlier passage, when it shifted Mars away from the earth) was about 20,000 years ago.

This might be the event "when mankind (in the early days of Seth and his star-watching priesthood (the royal line of Adam)) began to call upon the name of the Lord (Yahweh)" (Genesis 4:26). Something had terrified mankind in the heavens, which Seth's men began to study intently. Seth even received the nickname "Typhon." (Turner and Coulter 2001, p. 480)

It was 20,000 years ago that last ice age and the Great Arabian Desert began with abrupt suddenness, after ages of rain. For 80,000 years before that, it is thought, Arabia had been lush with rainfall. (Clapp 1998, pp. 220-221)

Something fundamental to earth climate had changed. We suspect that the earth's orientation or its orbit became disturbed by the close passage of something very massive, like Typhon.

Then, after 10,000 years of lifeless desert, Arabia suddenly bloomed again as the rains mysteriously returned. (Clapp 1998, p. 221) The lost civilization of Ubar then sprang to life in Arabia, only to be suddenly swept beneath the sands about 5,000 years later. (Clapp 1998, pp. 221, 230)

Now one might be forgiven for proposing an alternative to Typhon's passages as an explanation for one of these coincidences with its close approaches, but if we dismiss all three as un-related to Typhon, then we are clearly ignoring the obvious: Arabia's cycle of drought and rainfall precisely matches the times of Typhon's orbital returns.

Like a gigantic hour-glass, the sands of Arabia sift steadily across its desert, as if a grand cosmic countdown were being measured out, waiting patiently for Typhon's orbit to bring it back and restore the rains.

What we see happening in the great Arabian tableau is the twisting of the earth's surface by the mighty magnetic grip of Typhon. Whenever Typhon comes close enough, it moves the outer crust of the earth about, triggering great changes in climate for Arabia.

Not every passage does this, however. For some 80,000 years before Seth's time, if the rainfall record of that early era in Arabia is correct, Typhon did not send Arabia into a severe drought. But then, 20,000 years ago, Typhon came unusually close to the earth, and a great famine began in the Arabian Peninsula.

What went wrong? This is the time when Mars seems to have been suddenly disturbed as well. Could these things be connected?

If Mars became entangled with the earth and Typhon 20,000 years ago, it may have been the catalyst for a change in the orbit or orientation of the earth. We need not speculate on the details of that encounter. The result is clear enough: The earth became cold and icy, and Arabia became brutally dry.

Even the titanic encounter of 15,000 years ago (when Zeus/Mars fought Typhon for the throne of heaven and the exploding comets melted much of the ice) could not end Arabia's misery. It remained parched, and all but lifeless, as the civilization of Atlantis arose, and then sank in Noah's Deluge 11,600 years ago. Only when Typhon returned about 1,300 years after the Flood did Arabia suddenly blossom again.

In fact, it is well-established that the whole northern hemisphere began to warm and prosper at that time. The change was planet-wide, not just in Arabia. This a further sign that Typhon had affected the whole earth, not just Arabia.

But we are still left with a great mystery: Why were the ancients so insistent that planets like Mars and Mercury had been "born" in the age just before Adam?

Remember the myth of Ophion? He had entwined himself about someone that the Greeks called Eurynome, "Lady of the North," the equivalent of the Egyptian "Tefnut" and the Hebrew Tsephon-ette, who may have been the mother or an earlier ancestor of Adam and Eve. She would have been born in "Tevel," the previous creation age of the rabbis, the one before our own Adamic age, which is called "Heled." (Velikovsky 1950, 1967, pp. 49-50)

According to the Greek myth, when the great Serpent Ophion wrapped himself around Eurynome/Tefnut, she became impregnated with the planet gods. Soon after the Serpent Ophion disentangled himself from her, she "gave birth" to the planets. This seemingly took place not long before the age of Adam began, hence, near the end of the previous age, the age of Tevel, which might also be spelled Tefel, the age of Tefnut.

There is a legend about the last Queen of that age, a woman who could correspond to Tefnut. She went weeping about her land, because of a cruel famine. (Turner and Coulter 2001, pp. 418-419, 456-457) If this famine was linked to Typhon's pre-Adamic passage, it might indicate a cosmic cataclysm 25,000 years ago.

Now we can begin to see the dim outline of the ancient planetary upheaval that helped end the pre-Adamic age of Tefel. It was a passage of Typhon that plunged their world into famine and death. And somehow, during that passage, the great Serpent known as Ophion, which John Mark tied to Satan's Red Dragon, whom we earlier have identified with Typhon, became "entangled" with Eurynome/Tefnut, an entity large enough to give birth to the planets. How can this be?

The Bible tells the same story: The comparison is with something called "Bohu" (usually translated "void"), who is the "female" ("Behemoth") half of the prior condition of the earth before Adam's world. In other words, Bohu (emptiness and desolation) corresponds to the world of Tefel after it was destroyed by Typhon and left to starve.

So, the starving world of Queen Tefnut is called Bohu. And, when we compare her Egyptian attributes, she is sometimes identified with Hathor, a gigantic wild cow or ox who looks and acts much like Behemoth (*cf.* Job 40:15). (*Strong's* #3930 תוהמב = Desolation and Death)

She is sometimes confused with Tehom, the masculine revived "sun" at the start of the Adamic world. More correctly, she is his mate, the goddess Tiamat (note the feminine ending: Tehom-ette); Tiamat is said to personify the waters of the pre-Adamic Deluge that brought an end to the Famine of Typhon. Her body of primordial water is said to have been "slain" and divided into two halves to make the firmament above and the earth and sea below (Turner p. 466). This corresponds to the thick fog of the new Adamic world before God separated it into the clouds above and the sea below on the second day.

But Tiamat had a curious description in some Semitic traditions: She was said to be an evil winged seven-headed red dragon, who became set on revenge when her mate was cast down: She equipped her son, Kingu (Cain-u) with horrible weapons and sent him off to make war on her enemies. Tiamat is also known as "Ba-au," which is the same as Bohu. (Turner and Coulter 2001, pp. 93, 466)

Note how her story matches the Revelation story of Lilith-Behemoth, the False Prophetess, whose husband, the Beast Cain, has been wounded by Seth. Lilith rises up as a great two-horned beast (Behemoth) to make an image of her wounded Beast husband, an image of Cain (Kingu), whom we know as Horus.

Thus, Satan, the seven-headed Red Dragon has two twin offspring: Cain and Lilith, the Beast and the False Prophet. We know that the Beast has seven heads and is red like the Dragon. It makes sense that his twin sister, the False Prophetess, would also look like a seven-headed red dragon. Yet, just like her Bohu-Behemoth alter-ego, she can transform herself into "an angel of light" by appearing as if she has "two horns like a Lamb," in her "Marian" apparitions, yet Lilith is really a Dragon, (Revelation 13:11) seeking to behead Christians who refuse to worship her son: Horus. (*cf.* Revelation 13:15, 17:6, 20:4)

Likewise, the Medusa female head of the three-headed Gorgon, or Kraken Dragon was once beautiful, but now is a hideous monster, the Night-Hag "turning men to stone," hypnotically paralyzing them to extract their seed to restore her beauty briefly. Perseus, who slays Medusa, represents David, the heir of her ancient foe, Seth.

Tiamat-Tefnut, then, represents the survivor of the pre-Adamic female lineage. Dragon-Queen Tefnut's current heiress is Lilith, the False Prophetess. Tefnut was perceived as symbolizing Tiamat/Bohu, the great primordial mist or fog that was split asunder to create heavenly and earthly waters. (Genesis 1:6). Tefnut and her husband Shu survived into the Adamic age, passing through this fog to bring Eve to Adam. (Genesis 2:24). Later, Shu and Tefnut leave with Cain and Lilith, wandering until they settle in Egypt. They tell their twin grandchildren stories of the pre-

Adamic world; had they not done so, the pagan traditions of that epoch would not have survived.

One of those tales told of how Mars and the other planets first became visible as the fog lifted, following years of misty gloom in the wake of the sun going dark. The now visible stars began to serve to mark the signs and the seasons. (Genesis 1:14-15) It was as if the fog Tiamat had somehow "given birth" to the planets. Maybe it was only a bed-time story for Tefnut's grandchildren, a way to poetically embellish her tale, but it was remembered.

But there was also an earlier, scarier tale. It told of how Typhon had come by, wrenching the earth around, leaving the world devastated and starving. Caught up in this story was the birth of Mars, a planet not seen before. (Turner and Coulter 2001, p. 175) It had apparently been captured by Typhon from a far place and was dragged into the earth's vicinity.

If one might hazard a guess, Mars could have been plucked from an orbit around Saturn or Jupiter. It would not have been noticed at that distance. It must have had a wide orbit to have been drawn off by Typhon. How could a planet like Mars get into such a tenuous and distant orbit?

The original exploding planet in the asteroid belt may have had its own moons, one of which was Mars. The explosion of the parent planet may then subsequently ejected Mars out into space. This idea has been put forward by author Richard C. Hoagland, among others, and it's as good as any right now. (Hoagland 1987, 1992)

But at some point, Mars had to find a stable orbit. I think when it first was kicked out of the asteroid belt with the added energy of the exploding parent planet, Mars could have been boosted into orbit around Jupiter or Saturn. But that orbit must have been on the outer fringe of their pre-existing moons. Such an outside orbit would have made it easy for Typhon to grab it and carry it back down into earth's vicinity.

Like a big black cat dropping a dead robin on our doorstep, Typhon deposited the red planet uncomfortably close to earth 25,000 years ago. The climatic disruption and earthquake activity stimulated by the perilous passages of Mars could have left millions of dead. The twisted surface of the earth shut down the old rain cycle. The world starved.

These events keep echoing down the ages like a repeating nightmare. Typhon is a far bigger predator than expected, capable of batting Mars around like a helpless bird. Spanning the heavens, Typhon will soon bring Mars to our doorstep once again. But as a predator torments its prey to drive it into hormone-drenched exhaustion, Typhon will extract terror from Mars until it disintegrates before our eyes. (Revelation 1:8, Ezekiel 28:16-19)

There are now more and more fireballs coming all the time. The rate of increase in fireball sightings has itself continued to increase and may even be accelerating faster than we had predicted in our chart in Chapter Seven.

Some may assume that an increase in the fireball rate means that the gravitational mass that is creating the clustering must be getting close (that is, Typhon). Not so. This is actually a sign that we are still in the early phase of the fireball bombardment.

Let me explain. The thickest part of the fireball debris-field is nearest Typhon. The debris has been falling on us at an ever-increasing rate. This is expected as the cloud of material first enters the vicinity of the earth. Fireballs arrive at an ever-escalating pace as we get deeper into the debris cloud. But when Typhon itself passes, the rate of increase will drop to zero, the bombardment having reached its peak intensity. Prior to that, the rate of increase must begin to slow down. But since the rate seems to be continuing to accelerate, Typhon must still be far away. It's not all "good" news: It means Typhon may be even larger (and deadlier) than we had assumed.

Chapter Bibliography

Clapp, Nicholas. 1998. *The Road to Ubar: Finding the Atlantis of the Sands.* New York, New York: Houghton Mifflin Company.

Frazer, Sir James G., and Theodor H. Gaster. 1964. *The New Golden Bough: A New Abridgement of the Classic Work by Sir James George Frazer.* New York: Mentor Books.

Hoagland, Richard C. 1987, 1992. *The Monuments of Mars.* Berkeley, California: North Atlantic Books.

Howell, Elizabeth. 2018. Shoemaker-Levy 9: Comet's Impact Left Its Mark on Jupiter. January 24. Accessed April 14, 2019. https://www.space.com/19855-shoemaker-levy-9.html.

Phillips, Tony. 2013. *Collision Course? A Comet Heads for Mars.* March 27. Accessed April 14, 2019. https://science.nasa.gov/science-news/science-at-nasa/2013/26mar_marscomet.

Sitchin, Zecharia. 1998. *The Cosmic Code.* New York: Avon Books.

Turner, Patricia, and Charles Russell Coulter. 2001. *Dictionary of Ancient Deities.* 1st. New York, New York: Oxford University Press.

www.ingramcontent.com/pod-product-compliance
Lightning Source LLC
Chambersburg PA
CBHW060824170526
45158CB00001B/76